Yingzhen Liu

Design of a superconducting DC wind generator

Karlsruher Schriftenreihe zur Supraleitung Band 022

HERAUSGEBER

Prof. Dr.-Ing. M. Noe
Prof. Dr. rer. nat. M. Siegel

Eine Übersicht aller bisher in dieser Schriftenreihe erschienenen Bände finden Sie am Ende des Buches.

Design of a superconducting DC wind generator

by
Yingzhen Liu

KIT Scientific Publishing

Karlsruher Institut für Technologie
Institut für Technische Physik

Design of a superconducting DC wind generator

Zur Erlangung des akademischen Grades eines Doktor-Ingenieurs von der KIT-Fakultät für Elektrotechnik und Informationstechnik des Karlsruher Instituts für Technologie (KIT) genehmigte Dissertation

von M. Eng. Yingzhen Liu

Tag der mündlichen Prüfung: 26. Januar 2018
Referent: Prof. Dr.-Ing. Mathias Noe
Korreferent: Prof. Dr.-Ing. Martin Doppelbauer

Impressum

KIT Scientific Publishing

Karlsruher Institut für Technologie (KIT)
KIT Scientific Publishing
Straße am Forum 2
D-76131 Karlsruhe

KIT Scientific Publishing is a registered trademark
of Karlsruhe Institute of Technology.
Reprint using the book cover is not allowed.

www.ksp.kit.edu

Print on Demand 2020 – Gedruckt auf FSC-zertifiziertem Papier

ISSN 1869-1765
ISBN 978-3-7315-0796-3
DOI 10.5445/KSP/1000083023

Acknowledgement

This thesis was written at the Institute for Technical Physics at Karlruhe Institute of Technology and it cannot be finished without the help of my colleugues.

I would like to thank my supervisor Prof. Dr.-Ing Mathias Noe, who provides me with the opportunity to pursue my PhD in Karlsruhe Institute of Technology. His continuous support, advice and insight have helped me to reach a higher research level. I highly appreciated the constructive feedback and helpful guaidance given by Prof. Noe at a regular meeting evey two to three weeks during my whole PhD period. In order to ensure the scientific quality of my work, Prof. Noe also encourages me to participate in international conferences, workshops and seminars, which benefit me a lot.

My special gratitude goes to my second referee Prof. Dr.-Ing Martin Doppelbauer for his useful lessons, advice and discussions on electric machines, and the excellent and professional environment he offered to study the iron material properties. Specially, I would like to thank Prof. Doppelbauer for his scientific input and linguistic improvements, that helped a great deal to finish the final version of this thesis.

I would like to thank my external referee Prof. Jean Lévêque for his scientific and practical comments which help me a lot to improve my thesis. Special thanks go to Rainner Gehring, Holger Fillinger, Hong Wu, Johann Wilms, Jörg Brand, Uwe Walschbuger, Andrej Kudymow, Bernd Ringsdorf, Anna Kario, and Fabian Schreiner. Many measurements could not have been performed, if not for their outstanding, technical support and their motivation to repair or replace broken or damaged equipment. I would also like to thank Francesco Grilli and Victor Manuel Rodriguez Zermeno, for their support and guidance on ac loss calculation of superconducting coils.

My thanks will also go for the entire Supra group for many on- and off-topic discussions, and social activities, which have always been a recreating highlight during a busy week.

Finally, I would like to express my sincere gratitude to my parents, Guangshan Liu and Liuduan Fan, to my brother and sister, Kailong Liu and Jingzhen Liu, and to my lovely husband, Jing Ou. I would never have made it this far if not for the support of my family, to whom I shall ever be indebted.

Karlsruhe, Nov.11, 2017
Yingzhen Liu

Zusammenfassung

Offshore-Windenergie ist eine zunehmende Quelle erneuerbarer Energien. Ein erfolgversprechender Weg zur Verringerung der Stromgestehungskosten dieser Erzeugungsart besteht in der Entwicklung großer Windparks und Turbinen. Neben der im Jahr 2015 durchschnittlich erreichten Windturbinengröße von 4,2 MW existieren auch bereits Turbinen mit Kapazitäten von 6 – 8 MW. Turbinen mit höherer Nennleistung sind in der Entwicklung und gelangen zukünftig zur Anwendungsreife. Der Trend hin zu höheren Nennleistungen und einer größeren Anzahl von Offshore-Anlagen erfordert Innovationen im Bereich der Windturbinen. Diese wiederum setzen ein geringeres Gewicht der Turbinen, geringere Kosten, kleinere Maße, höhere Wirkungsgrade und eine höhere Zuverlässigkeit voraus. Aufgrund der hohen Stromtragfähigkeit und der nicht vorhandenen Gleichstromverluste der Supraleiter lässt sich ein beachtliches Gewicht-Volumen-Verhältnis bei hohem Wirkungsgrad des supraleitenden Generators erreichen. Darüber hinaus ist die Gleichstromübertragung für Offshore-Windparks in erster Linie aufgrund des gesamtwirtschaftlichen Vorteils in Anbetracht ihrer Lage weitab des Festlands in den Vordergrund gerückt. Als mögliche technische Lösung stellt die vorliegende Abschlussarbeit in diesem Zusammenhang ein auf supraleitenden Gleichstrom-Windgeneratoren und Gleichstrom-Kabeln basierendes System zur Gleichstromerzeugung und –übertragung vor. Diese Lösung ermöglicht sowohl den Betrieb eines extrem leistungsfähigen und kompakten Generators als auch den Einsatz eines neuartigen und ebenfalls sehr leistungsfähigen Gleichstromanschlusses.

Die vorliegende Arbeit konzentriert sich auf eine Machbarkeitsstudie und die Entwicklung des supraleitenden Gleichstrom-Windgenerators. Es wird ein Entwurf des Generators entwickelt, der nach den vorgegeben Kriterien, Kosten, Masse, Volumen und Effizienz optimiert werden kann. Zum Verknüpfen des elektromagnetischen Designs und der mechanischen Auslegung mit den Eigenschaften der supraleitenden Bänder und Eisenwerkstoffe werden alle notwendigen analytischen Gleichungen abgeleitet. Zur Erhöhung der Auslegungsgenauigkeit werden die analytischen Gleichungen zur Berechnung der Flussdichteverteilung im supraleitenden Gleichstromgenerator durch Finite-Elemente-Analysen verifiziert. In die Massenberechnung werden neben den aktiven Bauteilen auch die inaktiven Konstruktionswerkstoffe einbezogen. Diesem Auslegungsverfahren folgend wird die Entwicklung eines supraleitenden 10 kW-Gleichstrom-Demonstrationsgenerators beschrieben. Dabei wird auf die Verluste des Demonstrators und seiner Kommutierung, seines Drehmoments und seiner Leistung bei unterschiedlichen Windgeschwindigkeiten eingegangen. Im Vorfeld der Entwicklung des Demonstrators werden die Eigenschaften von Schlüsselkomponenten wiesupraleitenden Bändern und Eisenwerkstoffen getestet und charakterisiert. Zur Ermittlung des möglichen Potentials großer supraleitender Gleichstrom-Windgeneratoren wird ein supraleitender 10 MW-Generator konzipiert und mit konventionellen Synchrongeneratoren verglichen. Die vorliegende Arbeit erörtert darüber hinaus auch die Einsparung der HTSL Bandleitermenge durch Optimierung des Außendurchmessers des Rotors, der Polpaarzahl und der Höhe der supraleitenden Spule. Diese Maßnahmen tragen zu einer wettbewerbsfähigeren Alternative zu konventionellen Generatoren bei.

Abstract

Offshore wind energy has received a lot of interest as one important renewable energy source. One promising way to reduce the Levelized Cost of Electricity (LCOE) of offshore wind energy is by developing large wind farms and turbines with large ratings. The average wind turbine size has reached 4.2 MW in 2015 and turbine sizes of 6-8 MW have already been seen in the wind market. Even larger turbine sizes are managing to pave their way from studies to market. The trend towards larger ratings and more offshore installations asks for innovations in power generation, which requires lower weight and cost, smaller size, higher efficiency and reliability. Due to the high current-carrying capability and no DC losses of the superconductors, superior power to weight/volume ratio with high efficiency of a superconducting generator can be achived. Moreover, direct current (DC) transmission has been put forward for the offshore wind farms mainly due to the overall economic benefit, as they are located far away from the land. Hence, this thesis introduces a DC generation and transmission scheme which consists of superconducting DC wind generators and superconducting DC cables as a possible technical solution. This enables a highly efficient and compact generator and in addition a new and also very efficient generator connection scheme at DC.

The work presented in the thesis focuses on the feasibility study and design of a superconducting DC wind generator. In part, an optimisation method will be developed by taking superconducting tape length (cost), mass, volume, and efficiency into a simplified objective function. All necessary analytical equations will be derived to connect the electromagnetic design and mechanical design with properties of the superconducting tapes and iron materials. To increase the design accuracy, analytical equations to calculate flux density distribution in the superconducting DC generator will be verified by finite element analysis. Not only the active parts but also inactive structural materials will be included in the mass calculation. Based on the design method, the design of a 10 kW superconducting DC generator demonstrator will be described. The losses of the demonstrator and its commutation, torque and efficiency at different wind speeds will be addressed. As first steps towards the demonstrator, properties of key components, superconducting tapes, iron materials and a superconducting coil, will be tested and characterized. Moreover, a preliminary test of a superconducting coil at 77 K will be completed. In order to identify the potientials that a large scale superconducting DC wind generator could offer, a 10 MW superconducting DC generator will be designed and a comparison with conventional synchronous generators will be made.

Table of Contents

1 Introduction

1.1 Motivation and scope of the work

Wind energy, as a clean and renewable energy, is now being widely developed to reduce carbon dioxide production. According to a report by the Global Wind Energy Council[1],[2], the cumulative installation in 2015 is about 433 GW, while the newly installed capacity is 63 GW in 2015. The estimation of the global installed wind capacity from the year 2015-2050 will be increased from 433 GW to 3984 GW [1],[2]. The need for wind energy motivates generators with larger rating, lower cost, lower weight, and higher reliability. One solution is the direct-drive generator concept, such as a permanent magnet generator and superconducting generator. The direct-drive system simplifies the drive train. When referring to weight, volume, and cost, superconducting generators are superior to permanent magnet generators for wind turbines with rated power of 8 MW or more according to a report from the American National Renewable Energy Laboratory [3]. The reason lies in the high current carrying ability of the superconductor, which can be used to produce higher magnetic flux density. As the capacity of the generator is proportional to the product of the current per circumferential length and flux density, the generator can achieve higher rating by employing superconductors. In other words, superconductors can contribute to a more compact and lighter weight design than conventional generators. American Superconductor (AMSC) has designed a 10 MW superconducting direct-drive generator with weight of about 160 tons, which is roughly about 50% of that of a permanent magnet generator [4]. Moreover, there is no DC resistance when DC current is applied to the superconductor and the resulting losses in the generator are reduced. The efficiency of the generator can reach as high as 98%, which is impossible to be achieved by conventional generators. Hence, superconducting direct-drive generators are promising candidates for 10 MW class wind generators.

However, in a superconducting synchronous generator, the cryostat and the superconducting rotor are rotating with the shaft and a transfer coupling, also named as rotary joint, is needed to transfer cryogenic coolant from stationary supply to rotary cryostat. To avoid that and simplify the cryostat system, the superconducting coils are better to be stationary. Correspondingly, superconducting DC generator can meet the requirement. Moreover, as many of the offshore wind farms are located far away from the land, a direct current collection and transmission topology for the grid has been put forward for future offshore power transmission [5]-[8]. Superconducting DC generators and superconducting DC cables can be employed to enable a new DC power generation, collection and transmission scheme. A brief sketch of this proposed scheme is shown in Fig. 1.1. By applying the superconducting cable, only medium voltage DC/DC converters are needed and the voltage step up to high levels are not required. Another advantage of the superconducting DC generator is that no rotating exciter is needed for the superconducting coils and only DC current flows in the superconducting coils. The main drawback is the commutation of the DC generators. The common ways to solve it are by introducing interpoles, shifting brushes, and using compensation windings. A detailed design of a 10 MW superconducting DC motor including commutation was made in 1994 [9].

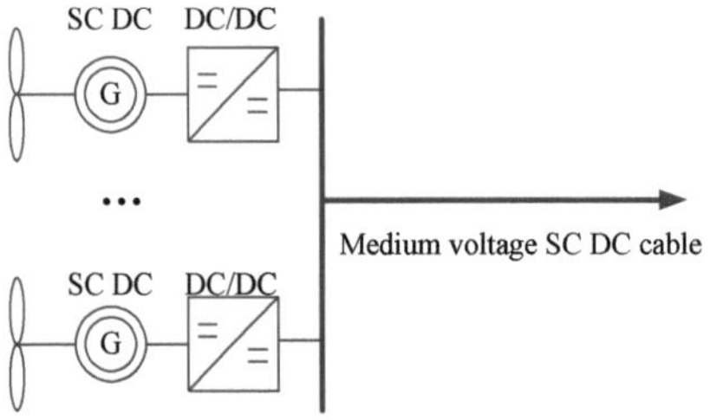

Fig. 1.1: Brief sketch of a superconducting DC generation and transmission scheme.

This work proposes a superconducting DC generator for offshore wind turbines to improve the overall efficiency and reduce energy cost. The superconducting coil is used in the stator to allow only DC current in the superconductor and a stationary refrigeration cooling system. To confine the flux density from going outside the generator, a stator iron core is used. A dewar and a thermal shield is employed to maintain the cryogenic working temperature at about 30 K. There are coil former and coil support structures to provide mechanical support and thermal conduction for the superconducting coils inside the dewar. In order to avoid high AC losses of the superconducting tapes and complex cryogenic cooling system, airgap copper windings are used in the rotor instead of superconducting tapes. The rotor iron core is introduced to reduce the magnetic reluctance.

The main objectives of the work are:

- Develop necessary theoretical background and equations for the superconducting DC generator. Integrate the design equations into an optimization method based on design objectives including mass, volume, efficiency and superconducting materials cost.

- Develop an analytical electro-magnetic and mechanical model to calculate the magnetic flux density in a superconducting DC generator and estimate the structural mass of it. Validate the calculated results by finite element software.

- Design a superconducting DC generator demonstrator with the output power of 10 kW. Validate the performance of the demonstrator by finite element software.
- Conduct measurements of the key components in the demonstrator, including tests of the critical current of the superconductor and superconducting coil, *B-H* curves and losses of the silicon lamination sheets at low temperature.
- Compare the conceptual design of a 10 MW superconducting DC generator by applying the obtained results of the measurements with superconducting and permanent magnet synchronous generators at the same rating.

The rest of this chapter is furtheron focused on the state of the art of the 10 MW class wind generators. Different generator concepts are reviewed and compared through a thorough literature study. The fundamentals of technical superconductors and permanent magnets are summarized. Moreover, the current status of the wind energy is also described.

Design equations of the superconducting DC generator are presented in Chapter 2. To design a superconducting DC wind generator, there are many geometric dimensions and material characteristics. In the design process, a large input data for the variables is necessary in order to find the optimal design which satisfies the given design specifications and objectives. In order to take all the variables into consideration, an optimization method is proposed. By using normalization and weighting factors, the optimization objective function is built based on weight, volume, cost and efficiency and superconducting tape length, which can indicate the cost of the generator.

Chapter 3 describes an analytical design model based on Poisson and Laplace Equations to calculate the magnetic flux density in a superconducting direct-current generator. The magnetic flux density generated by both superconducting coils and airgap armature winding is analyzed. The maximum perpendicular and parallel flux density in the superconducting coils are used to determine the tape operation current. The materials in the stator or rotor core are also taken into account. Furthermore, the model allows for the nonlinearity of the ferromagnetic materials and the operating relative permeability can be determined by an iterative algorithm. The results obtained by the model are compared with that of a finite element method.

To include the structural mass in the weight calculation, an analytical mechanical model is employed and explained in Chapter 4. Otherwise the optimal design with only active weight of the generator can be impractical. From a structural point of view, the support structure not only needs to be strong enough to withstand forces, but also maintains the air gap between the rotor and stator under rated load. For simplification, the rotor and stator is supported by a disc structure without any special weight reduction methods. After the derivation of the equations, finite element software is used to validate these equations.

By using the developed integrated electromagnetic-mechanical design method, a 10 kW demonstrator is optimized in Chapter 5. A superconducting coil is designed in detail and the normal operation performance of the demonstrator is analyzed. As there are different sources of losses in the demonstrator, loss analyses are conducted to calculate the efficiency. The AC loss of the superconducting coils and refrigeration losses are estimated. Furthermore, the no-load voltage and efficiency over a wide range of wind speeds are calculated and discussed.

At present, there is still a lack of experimental data for the key components of the superconducting generator. Hence, tests are carried out to analyze the basis of the materials and a double pancake superconducting coil in Chapter 6. The critical current of the superconducting tapes at different magnetic fields, temperatures as well as different angles to the magnetic field is studied. The magnetization behavior and losses of the silicon lamination sheets at both room and cold temperature are also measured. Finally, a superconducting coil used in the demonstrator is tested.

Future prospects of the superconducting DC generator are explored in Chapter 7 with a conceptual design for 10 MW. The aim of the 10 MW design is to study the feasibility and advantages of the superconducting DC generator. Moreover, a 10 MW superconducting and permanent magnet synchronous generator is designed respectively. Then the DC generator is compared with them in terms of torque density, size, and efficiency.

The results of this work are summarized in Chapter 8 with an outlook on future R&D work.

1.2 Status of 10 MW class wind generators

1.2.1 Wind energy

In Europe's perspective, renewable energy will provide 45% and 60% electricity by 2020 and 2030 and the contribution of solar photovoltaic and wind energy will rise from currently about 10% to 30% by 2030 [10]. Moreover, the governments worldwide are paving the way for the development of wind energy, such as carrying out new policies and legislations. As a result, the wind market continues growing. The estimation of the global installed wind capacity from the year 2000-2050 is shown in Fig. 1.2 [1],[2]. It can be noted that the annual installation keeps increasing every year except the year 2013 and by 2050 the global cumulative wind capacity is expected to be close to 4000 GW. The wind market reached 148 GW in Europe by the end of 2015. In particular, in Europe the annual offshore market has more than doubled compared to 2014 in 2015[1].

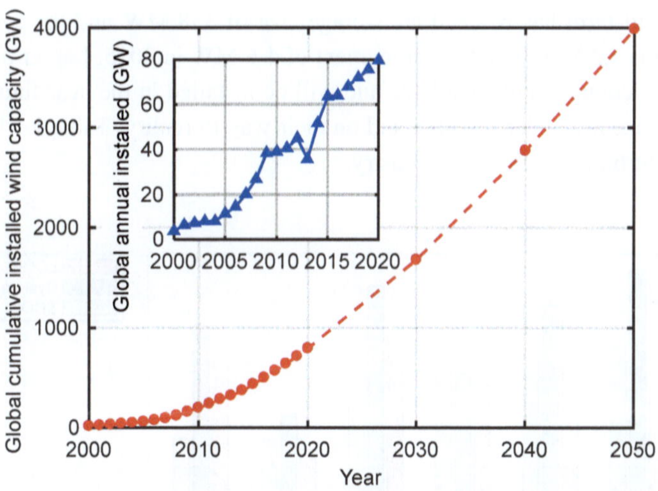

Fig. 1.2: Estimation of the global installed wind capacity from the year 2000-2050 [1],[2].

Levelized Cost of Electricity (LCOE) is widely used to evaluate energy sources and investment. However, the LCOE only takes the capital and operating cost into account. In order to have a better estimation, system cost of electricity, including subsidies, grid costs and variability costs, needs to be considered. To have a more comprehensive assessment of different energies from the benefits of a society on a macro scale, Siemens applies a concept of Society's Cost of Electricity (SCOE), which take the social costs, economic impact and geopolitical impact into consideration [11]. The costs of different electricity sources are compared for the installations in the UK in 2025, as shown Fig. 1.3 [11]. Up to date, the fossil fuels provide lower cost of energy when LCOE is considered. Based on SCOE, the cost of offshore wind energy changes remarkably and will offer the cheapest cost of energy, together with onshore wind. The target price, aimed at the industry, for wind energy by 2020 is €100/MWh [12]. It can be concluded that the offshore wind will be a pillar for future energy supply.

Even though the risk and capital cost of the construction of offshore wind are higher than onshore, the offshore wind yield is more stable and predictable [13],[14]. To catch up with the national maritime spatial planning and to capture and benefit from better resources at sea, offshore wind farms are aiming at sites further from land and deeper into water. The average water depth and distance to coast of offshore wind farms is 27.1 m and 43.3 km, respectively [15]. To reduce the LCOE of the offshore wind, developing large wind farms and turbine sizes is a promising solution. Based on rapid technology development, the average offshore wind farms and turbine sizes have been increasing in the last 25 years, as shown in Fig. 1.4 and Fig. 1.5 [15].

The offshore wind farm has reached an average size of 338 MW and the average wind turbine installed is 4.2 MW with the deployment of 4-6 MW in 2015. Larger turbine sizes with 6-8 MW are commercially available and will be installed in the near future. 10 MW and even larger size are under research and on their way to reality. Table 1.1 lists the top 10 largest wind turbines available in industry.

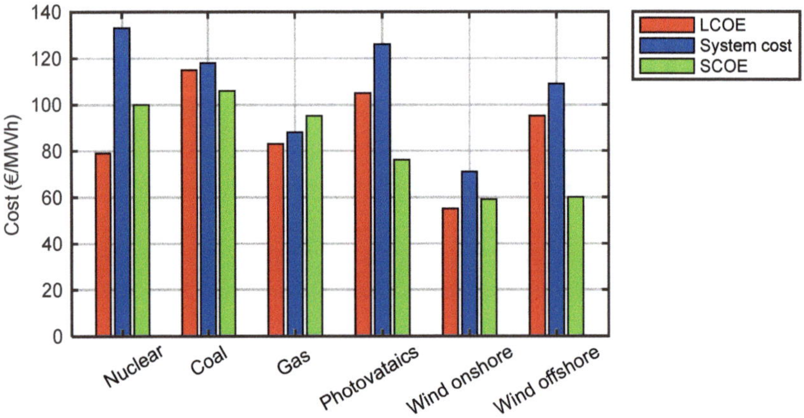

Fig. 1.3: Costs of different electricity sources for installations in the UK in 2025 [11].

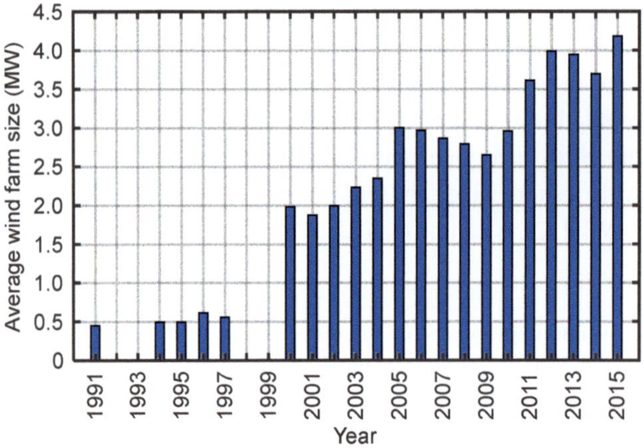

Fig. 1.4: Average offshore wind farm size from the year 1991-2015 [15].

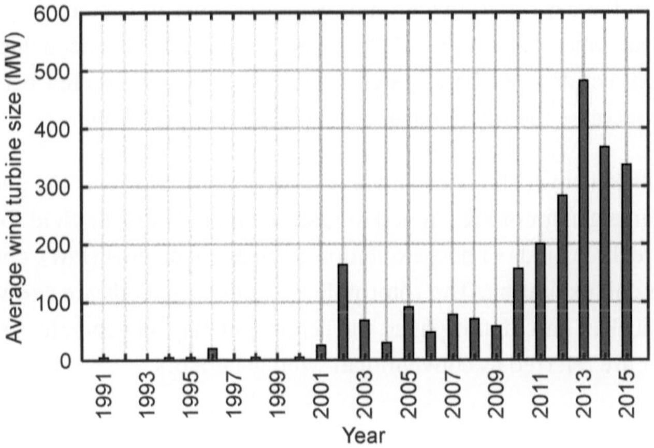

Fig. 1.5: Average offshore wind turbine size installed from the year 1991-2015 [15].

Table 1.1: Top 10 largest wind turbines which are commercially available

Name	Power rating	Drive train	Generator type	Rotor diameter	Ref.
MHI Vestas V164-8.0 MW	8 MW plus	Medium speed geared	PMSG	164 m	[16],[17],[18]
Adwen AD 8-180	8 MW	Medium speed geared	PMSG	180 m	[17],[19],[20]
Enercon E126	7.58 MW	Direct-drive	EESG	127 m	[17],[21]
Siemens SWT-7.0-154	7 MW	Direct-drive	PMSG	154 m	[17],[22],[23]
Ming Yang SCD	6.5 MW	Medium speed geared	PMSG	130 m	[17],[24],[25]
Senvion 6.2 M152	6.15 MW	High-speed geared	DFIG	126/152 m	[17][26],[27]
GE Haliade	6 MW	Direct-drive	PMSG	150.8 m	[17],[28]
Sinovel SL6000	6 MW	High-speed geared	DFIG	155 m	[17],[29]
Guodian United Power UP6000-136	6 MW	High-speed geared	DFIG	136 m	[30]
Siemens SWT-6.0-154	6 MW	Direct-drive	PMSG	154 m	[31]

The wind turbine system consists of several components to convert wind energy into electricity. For a direct-drive wind turbine system the main components are generators, machine housing, yaw drive, e-module, hub, rotor blades, tower and foundation [32]. The rotor blades and hub are used to capture the wind energy and convert it into rotational energy. The generator and machine housing then turn the rotational energy into electricity. After that, the e-module can transform the electricity to the grid. The yaw drives are needed to automatically turn the nacelle to the wind direction in order to capture maximum power. The tower and foundation are employed to support and bear all the mechanical load of the turbine. In non-direct-drive turbine systems, gear boxes are connected between the hub and generator to increase the rotational speed and decrease the torque.

A key component to convert the mechanical energy into electricity is the generator. Generally, the poles in a generator are excited to generate main magnetic field for energy conversion. There are three kinds of pole excitation: permanent magnet (PM), copper coil, and superconducting coil excitation. The rotor pole excitation methods are shown in Fig. 1.6, and δ is the physical air gap. In this paper, the permanent magnet generator and copper coil excited generator are referred as conventional wind generators.

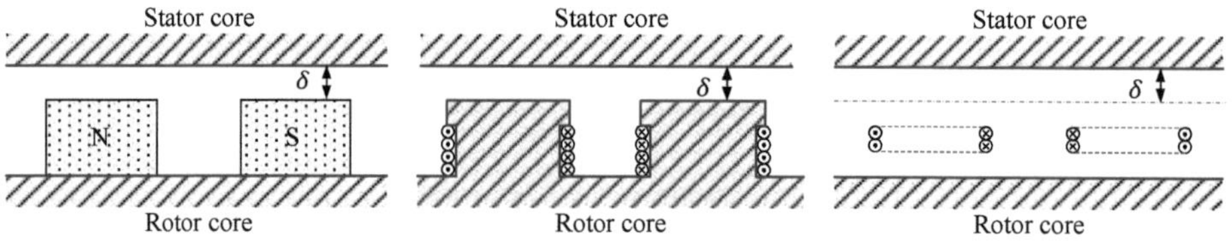

(a) Permanent magnet excitation (b) Copper coil excitation (c) Superconducting coil excitation

Fig. 1.6: Different pole excitation methods.

1.2.2 Copper coil excited wind generators

To help to reduce the reluctance in the magnetic circuit of the copper coil excited generators, ferromagnetic materials are employed as iron teeth, poles and pole shoes, and iron cores. In the field coil winding, the current density in copper winding, for example, in a salient pole synchronous generator is 2-4 A/mm^2, as shown in Table 1.2. With direct water cooling, the current density can be increased to 13-18 A/mm^2 and 250-300 kA/m [33].

Table 1.2: Current density of the field winding [33]

Generator types	Asynchronous	Salient pole synchronous	cylindrical synchronous	DC
Current density (A/mm^2)	3-8	2-4	3-5	2-5.5

There are two types of generators that are excited by copper coils: the doubly fed induction generator (DIFG) and electrically excited direct drive synchronous generator (EEDG), as illustrated in Fig. 1.7 and Fig. 1.8. The DIFG operates together with a high speed geared drive train, namely, a three-stage gearbox, which can turn the low rotational speed into high speed and decrease torque. The DFIG wind turbine system is illustrated as in Fig. 1.9.

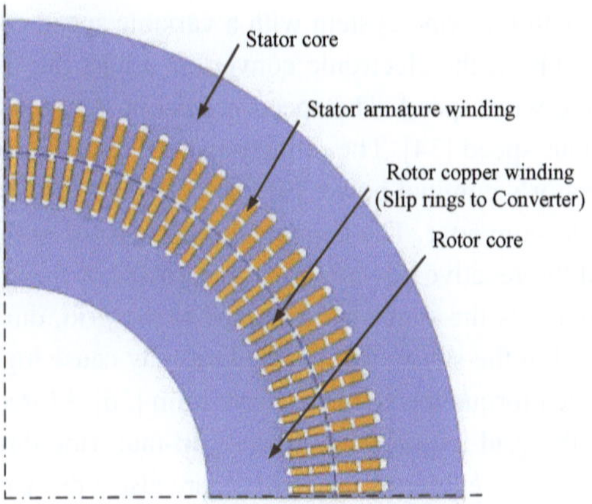

Fig. 1.7: Sketch of a doubly fed induction generator.

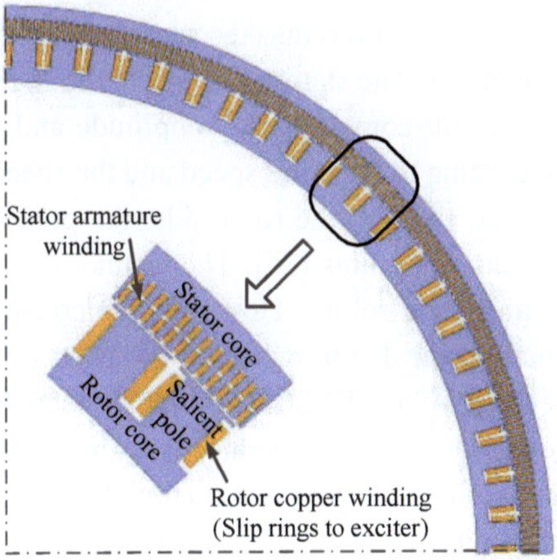

Fig. 1.8: Sketch of an electrically excited direct drive synchronous generator.

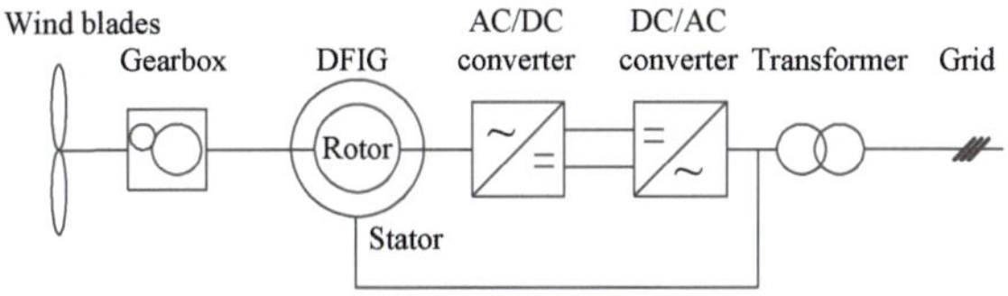

Fig. 1.9: Sketch of a DFIG wind turbine system.

The stator of the DIFG is directly connected to the 50/60 Hz grid network. The rotor of the DIFG is connected to a partial power converter through slip rings and brushes or a rotatory transformer. To work with the wind system with a variable speed control concept, which is the trend in wind turbines, the electronic converters adjust the frequency of the rotor current according to the wind speed. The speed regulation range of the DIFG is around ±30% of the synchronous speed [34]. The rating power of the electronic converter is typically 30% of the generator's nominal power [35], which has an advantage in the size, cost and loss reduction of the converter. The connection of the rotor to the converter also provides a way to control the reactive power and active power of the system by pulse width modulation [5]. However, as the stator is connected to the grid, during a grid fault, high fault current is produced in the stator, which in return may cause high current in the rotor and converter and sudden torque loads on the drive train [36]. Moreover, it is a challenge for the DFIG to meet the grid requirement of the grid fault ride-through capability [37]. The losses and reliability of the three-stage gearbox are also a drawback.

Differently, the EEDG has no gearboxes and generates the electricity directly from the low rotational speed. The EEDG wind turbine system is illustrated in Fig. 1.10. The rotor uses DC current for excitation and is connected to an exciter by slip rings and brushes or a rotating rectifier. A separated exciter or a reduced scale of converter connected to the grid is necessary for the rotor excitation. The stator is connected to the grid network through a full power converter. With the full converter, the amplitude and frequency of the stator voltage can be controlled according to the wind speed and the speed range is wide, even at very low speed [37]. Moreover, through the rotor side, the pole flux can be adjusted to minimize loss in different operating points [36]. The technology to design, manufacture, and control the EESG is mature and robust. However, the converter capacity should be larger than generator nominal power due to reactive power. As a result, the loss, size and cost of the converter is a demerit. As the EESG runs at low speed, the torque it produces is very high to meet the rated power and the generator is large in size and heavy in weight. Furthermore, the large amount of copper and losses in both field and armature winding also deteriorate the benefits. The comparison of direct drive generator system and generator system with gearboxes have been summarized in literature [13],[38]-[43].

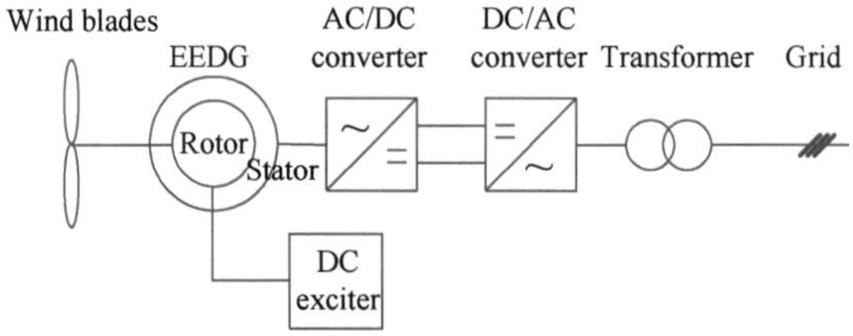

Fig. 1.10: Sketch of a direct-drive EESG wind turbine system with DC exciter.

1.2.3 Permanent magnet wind generators

1.2.3.1 Introduction of permanent magnets

In the past centuries, the performance of permanent magnets (PM) was remarkably improved. Many different types of PMs emerged, and they are widely used in electric and electronic devices from computers, appliances to medical equipment [44]. Because of its high power density and efficiency, the synchronous generators with PMs are being applied in wind turbines. At the moment, there are mainly four kinds of PMs being used: Ferrites, SmCo and FeNdB and the main properties of these PMs are summarized in Table 1.3.

Table 1.3: Main properties of permanent magnets

PM types	Ferrites	SmCo	NdFeB
Remanence	0.2-0.46 T	0.87-1.17 T	max. 1.43 T
(BH)max	6.5-41.8 kJ/m^3	143-255 kJ/m^3	max. 400 kJ/m^3
Electrical resistance	1010 uΩcm	1010 uΩcm	50-90 uΩcm
Temperature	max. 450 Celsius	max. 350 Celsius	80-230 Celsius
Price	cheap	expensive	medium

Ferrite magnet is the cheapest PMs [45]. Although the price of ferrite magnet is competitive, it is not the first option for wind turbine application as it has low remanence and energy product (BH)max. Its remanence is about 0.2-0.46 T and the (BH)max is about 6.5-41.8 kJ/m^3 [46]. Compared with other PMs, ferrite magnet has a much higher electrical resistance and Curie temperature. Its electrical resistance is about 1010 uΩcm. This makes it competitive in specific occasion as it will cause low eddy current loss.

In the 1960s, rare-earth PMs based on rare-earth-cobalt (R-Co) intermetallic RCo5 type compounds were discovered. Among them the SmCo5 PM has the highest coercivity and was recognized as the first generation of rare-earth high-performance magnets. Soon the second generation of rare-earth PM based on Sm2Co17 compound was discovered. These magnets not only have large coercivity and energy products, but also have excellent thermal stability [44]. SmCo5 PMs can work in temperatures up to 250 Celsius, while Sm2Co17 PMs can work at higher temperature up to about 350°C [47],[48]. However, SmCo PMs are more expensive. Compared with AlNiCo magnets, the remanence of SmCo is relatively low, and it is about 0.87-1.17 T. However, the (BH)max of SmCo is higher, it is about 143-255 kJ/m^3.

Due to the high price of Sm and Co, Co free magnets with similar characteristics became a research target. In 1984, Fe-based high-energy product magnets based on the ternary Nd2Fe14B phase were reported. Before 2011 the Nd2Fe14B was widely used for decades because of its excellent magnetic properties. In order to improve the performance at elevated temperature, Dy and Tb are widely used as doping elements in NdFeB magnets [44]. The more percentage of Dy in weight added, the higher the temperature NdFeB magnets could work at. At the moment, the remanence of NdFeB magnets can reach 1.43 T, and its (BH)max can reach 400 kJ/m^3. The working temperature of NdFeB magnets is relatively low, and the range is about 80-230 Celsius according to different grades. The temperature coefficient of induction and coercivity are relatively high. The electrical resistance of it is about 50-90 uΩcm [47].

1.2.3.2 Permanent magnet generator types

There are two drive-trains of the PM generator, one is with gearbox to increase the rotational speed to medium speed by a first stage gearbox, and the other is direct drive system, as illustrated in Fig. 1.11. The stator of the PM generator is connected to the grid through a full converter, similar to the EEDG. It has the merits and demerits of the drive train with full converter as aforementioned. A radial flux synchronous PM generator is illustrated as in Fig. 1.12. The armature winding in the stator can be a lap winding or a concentrated winding. Fractional-slot concentrated winding, widely adopted in the brushless PM machines, has shorter end winding length, higher slot fill factor and less number of slots [49]- [51]. As the coils are wound around one stator teeth, the manufacture is simpler than the overlapped winding. However, the concentrated winding always subjects to a lower winding factor than the lap winding [52],[53]. PMs are used in the rotor to produce the flux density and an iron core is employed to confine flux. There are different configurations of the rotor, as shown in Fig. 1.13. To obtain a high air gap flux density, the high performance PM is a good choice. However, the high performance PM is usually very expensive, and it still needs to be improved. For a specific PM, the thickness of the PM can be increased to improve its operating point, as well as the width. But it is better to increase the width than the thickness, as less PMs are needed to provide the same air gap flux density by increasing the width than the thickness. However, for the configurations as shown in Fig. 1.13(a) and Fig. 1.13(b) the total width of the PM is limited by the perimeter of the rotor. In that case, Fig. 1.13(c) can be employed to further improve the width.

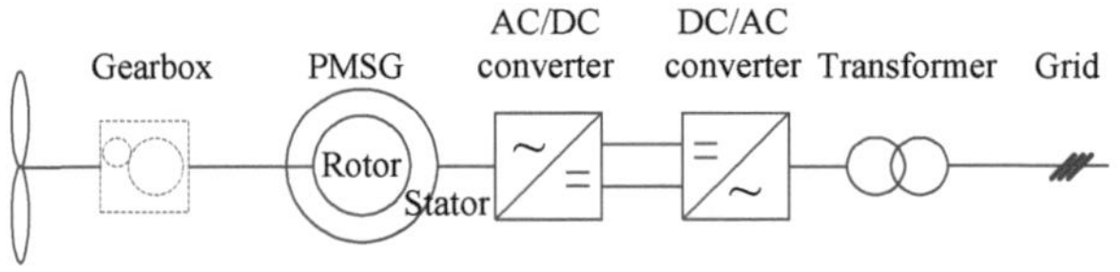

Fig. 1.11: Sketch of a PMSG wind turbine system.

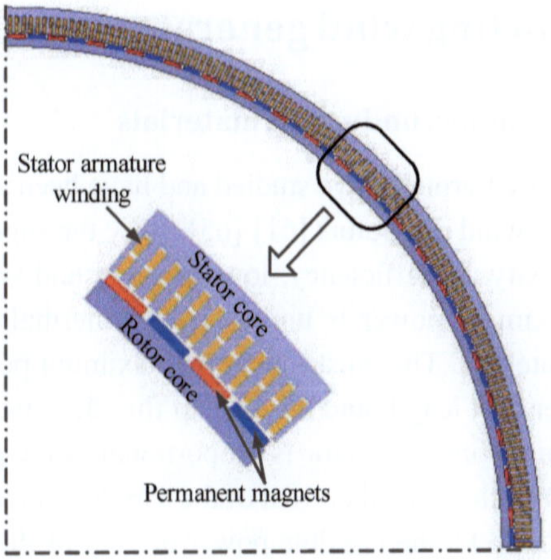

Fig. 1.12: Sketch of an electrically excited direct drive synchronous generator.

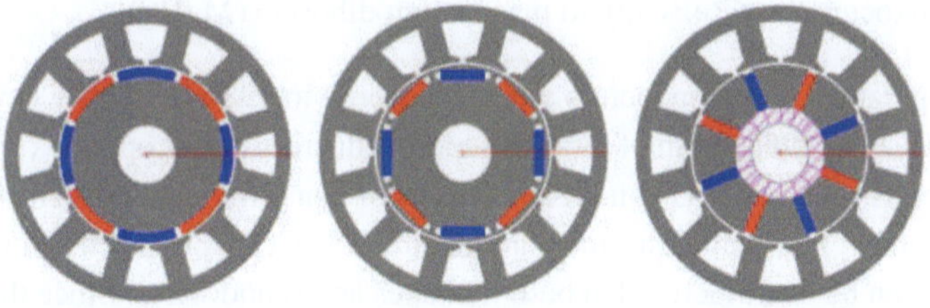

(a) Surface-mounted PM machine (b)Interior PM machine (c) Buried (spoke) PM machine

Fig. 1.13: Rotor configurations of radial flux PM generators.

One more advantage of the permanent magnet generator system is that there is no need for an exciter and it removes the slip rings and brushes, as well as the field losses, which lead to an increase in efficiency and reliability. However, the price of the permanent magnets is higher than copper, and the manufacture of permanent magnet is more difficult than copper.

Besides the most adopted radial flux topology of the permanent magnet generators, there are various topologies, such as axial flux and transverse flux types. Literature [54]-[58] compared and listed the advantages and disadvantages of radial, axial and transverse flux types and the project Upwind found that a single-sided single-winding flux-concentrating transverse flux PM generator was most promising [59]. The INNWIND project, from 2012-2017, is focused on innovative offshore wind turbines and components, and studies the pseudo direct drive permanent magnet generator, which is a non-contact magnetic gear integrated within a permanent magnet generator. It finds that this type has an extremely high torque density which eliminates any mechanical gearing and meshing teeth [60].

1.2.4 Superconducting wind generators

1.2.4.1 Introduction of superconducting materials

Many companies and research groups have studied and have been studying large-scale direct-drive superconducting wind generators [61]-[65], since the superconducting generator enables higher torque density and efficiency, lower weight and smaller size. In conventional generators, the maximum power is limited by the thermal heating and saturation point of ferromagnetic materials. This means that the maximum power is subjected to the circumferential current per unit length and the air gap flux density. Differently, the maximum power of a superconducting generator is proportional to the product of flux density and the critical current. Since the critical current of the superconductor is much higher even at a high external magnetic field, the possible power density of the superconducting generators in theory can be a factor of 10-100 better than the conventional generators with equal volume [66]. The superconducting materials investigated are varied from 1G superconductors, barium strontium calcium oxocuprate (BSCCO), 2G superconductors, rare earth barium oxocuprates (ReBCO), to magnesium diboride (MgB_2).

The high temperature superconductors were discovered in 1986 [67]. YBCO was found in 1987 and BSCCO in 1988. Both of the two superconductors have copper oxide planes that carry the currents. The BSCCO conductor consists of many filaments which are embedded in a silver matrix. The technique to manufacture BSCCO tapes is the powder in tube method and it can be manufactured in both flat tapes and round wires. Since there is a large portion of silver in the conductor, it has good mechanical properties and excellent thermal stabilization. However, it has a strong anisotropy, which leads to loss of current carrying capability in magnetic field, especially perpendicular field. It also suffers weak links, a significant amount of bad current paths. The main manufacture of this type of superconductor is Sumitomo Electric in Japan. The technical data provided by the supplier is summarized in Table 1.4 [68],[69].

Different from BSCCO, the ReBCO are formed as a multilayer coating on a flat substrate. The superconductor layer, which is very thin, typically 1 um thick, is deposited on the initial substrate, normally a nickel alloy. After this, a thin silver layer is added to protect the superconductor and to stabilize it both thermally and electrically. Then an outer layer of copper is applied to enable current transfer and stabilization and strand protection. The ReBCO tapes have better in field performance than BSCCO, which means that they can still conduct a large amount of current when exposed to magnetic field. Currently, there are many commercially suppliers all over the world. The tapes they produce differ in critical current, lift factor, geometry, stabilizer and price, etc. Table 1.5 lists the technical data from several manufacturers [69]-[76]. Both ReBCO and BSCCO can work at a wide range of temperature up to 77 K, which is the temperature of liquid nitrogen. In wind generators, the working temperature of the superconductors is selected to be 20 K to 40 K to enable a larger current density and they are cooled by conduction cooling.

In 2001, MgB_2 was discovered as a new type of superconductor [67], which has a critical temperature of 39 K. The operating temperature of MgB_2 in wind generators is chosen as 20 K. Similar to BSCCO, the MgB_2 superconductor is manufactured by the powder in tube method. It can be produced in many configurations, both in tapes and wires, which make it attractive for energy application. This powder in tube technique, together with the low cost of raw materials, makes MgB_2 cheap and easy to manufacture. However, the applied field to MgB_2 cannot be high, usually 1-2 T for applications. Otherwise, the critical current will drop dramatically. Commercially available MgB_2 from Columbus, Hitachi and Hypertech are summarized in Table 1.6 [69],[77]-[79].

Many advances have been made in superconductors, resulting in improvement of conductor quality and available length. The production capacity of 2G tapes reaches 1000 km/year. The price of the superconductors is depending on companies and the volume of the customers' order. The current price is around 100-1000 $/kA-m for ReBCO (using critical current at 77 K, self-field) and the price of the BSCCO is lower, which is in the range of 50-300 $/kA-m. Data released from IEA in ASC conference in 2016 shows that the ReBCO price can be reduced to below 25 $/kA-m in the year of 2030 [69], which is the threshold for large commercial market [80]. BSCCO will also decrease to the range of 10-50 $/kA- m in the year of 2030 [69], which will be higher than ReBCO, mainly due to the inevitable presence of silver matrix. The price of MgB_2 is below 25 $/kA-m (using critical current at 20 K, 1 T) and producers are aiming at further price target below 5 $/kA-m by 2025 to enlarge the market penetration [81]. To summarize, researches are conducted all over the world to increase the current carrying capability and to reduce the cost of the conductor.

Table 1.4: Technical data of commercially available BSCCO from Sumitomo Electric[68],[69]

	Type H	Type HT-SS	Type HT-CA	Type HT-NX	Type ACT	Type G
Width	4.3±0.2 mm	4.5±0.1 mm	4.5±0.1 mm	4.5±0.2 mm	2.8±0.1 mm	4.3±0.2 mm
Thickness	0.23 ±0.01 mm	0.29 ±0.02 mm	0.34 ±0.02 mm	0.31±0.03mm	0.32±0.02 mm	0.23±0.01 mm
Piece length	up to 1500 m	up to 500 m	up to 500 m	up to 500 m	--	--
Ic (77K, sf.)	170-200 A	170-200 A	170-200 A	170-200 A	60A,70A	170-200 A
Critical wire tension (RT)	80N	230 N	280 N	410 N	150 N	80 N
Critical tensile strength (77 K)	130Mpa	270 Mpa	250 Mpa	400 Mpa	270 Mpa	130 Mpa
Critical tensile strain (77 K)	0.02%	0.40%	0.30%	0.5%	0.40%	--
Critical double bend diameter(RT)	80 mm	60 mm	60 mm	40 mm	40 mm	80 mm
Critical current @20K	432A@2T, 383A@3T, 348A@4T	419A@2T, 371A@3T, 329A@4T	419A@2T, 371A@3T, 329A@4T	419A@2T, 371A@3T, 329A@4T	--	--
Critical current @30K	305A@2T, 240A@3T, 191A@4T	297A@2T, 234A@3T, 190A@4T	297A@2T, 234A@3T, 190A@4T	297A@2T, 234A@3T, 190A@4T	--	--
Critical current @40K	145A@2T, 74A@3T, 34A@4T	145A@2T, 74A@3T, 34A@4T	145A@2T, 74A@3T, 34A@4T	145A@2T, 74A@3T, 34A@4T	--	--
Production capability	1,000 km/year					30 km/year

Table 1.5: Technical data of commercially available ReBCO [69]-[76]

Supplier	Width	Thickness	Piece length	Ic (77K, sf.)	Critical tensile strength (77 K)	Critical tensile strain (77 K)	Critical double bend diameter(RT)	Lift factor	Production capability
Fujikura	4mm	--	300m	200A	--	--	--	--	--
	10mm	--	1km	467A	--	--	--	--	--
SuperOx	4mm	60-100um	50-300m	70-150A	770Mpa	0.45%	30 mm	2.45@20K,2T, 1.95@20K,3T,	--
	6mm	--		100-200A	--			1.67@20K,4T;	--
	12mm	--		250-500A	--			1.21@40K,2T, 0.96@40K,3T, 0.80@20K,4T;	300km /year
Sunam	4mm	--	1km	100 - 200 A	250Mpa	0.30%	30 mm	1.05@20K,2T, 0.7@20K,3.5T; 0.7@30K,2T, 0.45@30K,3.5T; 0.45@40K,2T, 0.3@40K,3.5T,	720km/year
	12mm	--	--	300- 600 A	--	--	--	--	--
SSTC	4mm	--	<=1km	80-200A	--	--	--	--	--
	4mm	--	--	100A	500Mpa	0.30%	60mm	2.2@30K,2T, 1.6@30K,3.5T; 1.5@40K,2T, 1.2@40K,3.5T,	--
Theva	11.8 - 12.2 mm	0.10 – 0.12 mm	up to 1km	360A	600Mpa	0.30%	60mm	--	150km/year
	12.0 - 12.5 mm	0.20 – 0.23 mm		360A	340Mpa	0.30%	60mm	--	
	12.0 - 12.5 mm	0.15 - 0.17 mm		360A	420Mpa	0.30%	60mm	--	
	11.9 - 12.25 mm	0.14 – 0.16 mm		360A	450Mpa	0.30%	60mm	-	
D-nano	4mm	0.1 mm	>50m	250A/cm (width)	150Mpa	--	10/30 mm	500A/cm(width)@30K,1T	planned 200 km /year
	12mm	0.1 mm				--			
Superpower	2mm	0.1 mm	100-300m	50A	--	0.45%	11mm	3.65@20K,2T, 2.88@20K,3T,	--
	3mm	0.1 mm		75A	--	0.45%	11mm	1.35@20K,4T;	--
	4mm	0.1 mm		100/140A	--	0.45%	11mm	1.89@30K,2T, 2.32@30K,3T,	--
	6mm	0.1 mm		150A	--	0.45%	11mm	2.96@30K,4T;	--
	12mm	0.1 mm		300A	--	0.45%	11mm	2.28@40K,2T, 1.75@40K,3T, 1.42@40K,4T;	--

	12mm	0.30 mm - 0.36mm	--	400-500A	200Mpa	--	95mm	--	--
	12mm	0.22 mm - 0.28mm	--	200-250A	200Mpa	--	70mm	--	
AMSC	12mm	0.18 mm - 0.22mm	--	205-350A	150Mpa	--	30mm	--	--
	4.24 - 4.55mm	0.36 mm - 0.44mm	--	70-100A,140-180A	200Mpa	--	35mm	--	--
	4.8mm	0.17 mm - 0.21mm	--	80-100A	150Mpa	--	30mm	--	--
STI	3 mm 4 mm 10 mm	--	--	250 to 500 A/cm	--	--	--	200-350A/cm @4.2K, 15T 4800A/cm @20K, 0T	--

Table 1.6: Technical data of commercially available MgB_2 [69],[77]-[79]

	Hitachi	Columbus superconductors	Hypertech
Dimension	1.5 mm O.D.	3mm×0.5 mm	0.7-0.9 mm O.D.
Piece length	300 m	1 to 5 km	1-4 km
Critical current @ 20 K	--	40A, 158A, 208A @ 2T; 60A,107A @3T; 13A,39A@ 4T	--
Engineering current density @ 20 K	>25A/mm^2 @3T	--	22-31 kA/cm^2 (nominal)@2T
Sheath	Monel	Ni Matrix	--
Production capability	--	3,000 km/year	--

1.2.4.2 Superconducting generator types

Since Prof. Heike Kamerlimgh Onnes discovered superconducting phenomenon in 1911, engineers were trying to apply superconductors into electrical machines. It was until mid-1960s when low temperature superconductors NbTi became into use that superconducting electrical machines became viable [82]. Between 1970s and 1990s, low temperature electrical machines gained a rapid development in research. For instance, Super-GM project in Japan, developed three rotors and a stator for 70 MW superconducting generators [83]. Then high temperature superconducting electrical machines gained a lot of attractions from research laboratories and industries, after high temperature superconductors Bi-2223, Bi-2212 and YBCO with higher transition temperature were available. Currently, the critical current of the YBCO is around 100-200 A with 4 mm width and 200-600 A with 12 mm width as shown in Table 1.5. Zenergy and Converteam initiated an 8 MW high temperature superconducting wind generator project in 2004 in UK [84]. Then AMSC started to design a 10 MW high temperature superconducting wind generator in 2007 [85]. In 2015, Ecoswing project announced to build a 3 MW class direct drive superconducting wind turbine, which is based on 2G wires [86]. This will be the first superconducting wind turbine in operation at a real wind tower. However, both 1G and 2G high temperature superconducting tapes are relatively expensive, though the costs are very likely to decrease in the future. Actually, the presence of AC losses has limited high temperature superconducting tapes to be applied in fully superconducting electrical machines. MgB_2 superconductor, having comparable low cost and relatively high operating temperature and low AC losses, seems to be promising in superconducting electrical machines. The engineering current density of MgB_2 can reach 130-300 A/mm^2 as listed in Table 1.6. A few companies began to design 10 MW fully superconducting wind generators in 2011 and 2013 [87],[88]. Besides, Suprapower project designed a 10 MW hybrid superconducting generator which employs MgB_2 coils to generate main field in the generator, as well as the INNWind. EU project [64],[89]. In addition, Windspeed project, aiming at the technical and economic feasibility study of a 3.6 MW superconducting generator, also used MgB_2 [90]. It should be pointed out that all the projects except the Ecoswing project and activities are only stay in designs, small demonstrators or subcomponents and no full-scale prototype has been successfully built and operated in real tower yet.

Different topologies have been studied in superconducting wind generators, such as axial, radial, and transverse flux topologies, DC and synchronous rotating generators etc. The different topologies with their advantages and disadvantages used in literature have been summarized in [91]. Many papers have investigated the performances, including steady state and transition state [92]-[95]. Among different topologies, the superconducting generator with superconducting field coils and copper armature winding, as illustrated in Fig. 1.14, has got the highest attention. The copper winding in the EEDG rotor is replaced by the superconducting tapes, accompanying with a cryostat. Inside the cryostat, there is a coil former which provides mechanical support and thermal conduction for the superconducting coils, while the Dewar and thermal shield as well as vacuum helps to maintain the cryogenic working temperature. A cryogenic system is necessary to cool down the superconducting tapes and to provide cold power. Stator iron core is usually used to confine the magnetic flux. The rotor iron core is introduced to reduce the magnetic reluctance and reduce the usage of superconducting tapes. The drive train of the superconducting synchronous generator is similar to that of the direct drive permanent magnet synchronous generators and no gearbox is needed.

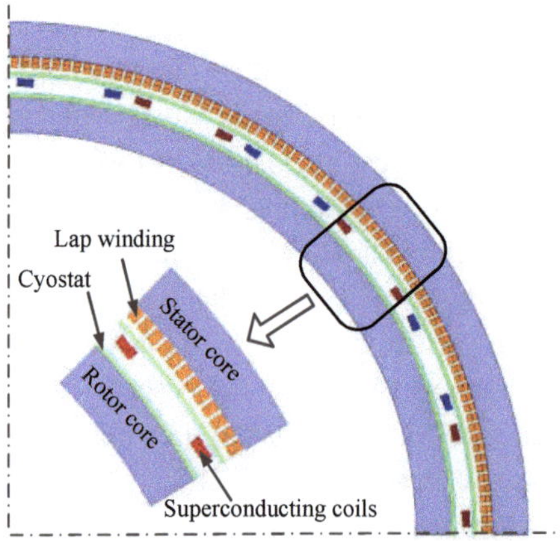

Fig. 1.14: Sketch of a superconducting synchronous generator with copper winding.

1.2.5 Comparison of different wind generators

In this work mainly direct drive generators are compared because they seem most promising for offshore generators with large ratings.

Mass and size of the generator are important due to the transportation and installation of the wind turbines. EU Up-wind project estimated the mass of the tower head and tower for 5, 10, 20 MW wind turbines, as shown in Fig. 1.15 [59]. In Fig. 1.15 the tower head mass includes the rotor mass and nacelle mass. Fig. 1.15 shows that the tower mass increase sharply when the power rating exceeds 10 MW. Reduction of tower head mass for offshore wind turbines can not only lead to decrease of the structural, foundation and installation costs, but also the transportation time [96]. For a 10 MW design with permanent magnets, the mass distribution of the tower head is estimated in Fig. 1.16 [80]. In the wind turbine, the generator mass is the heaviest, taking up 37% of the tower head mass. Hence, an effective way to reduce the tower head mass is to design a light weight generator. And transportation and installation also benefits from a small size generator.

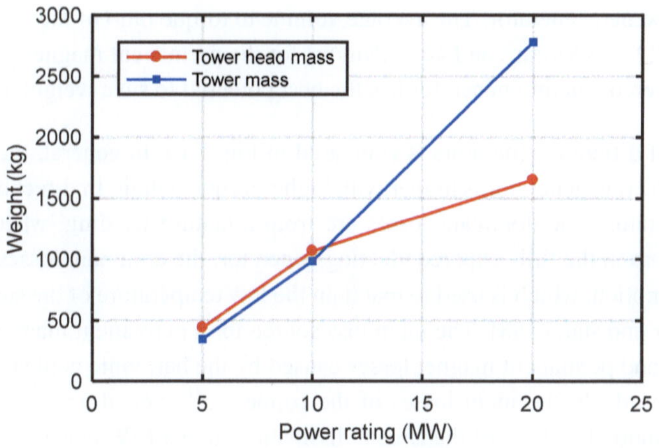

Fig. 1.15: Mass estimation for 5, 10 and 20 MW wind turbines of permanent magnet generators [59].

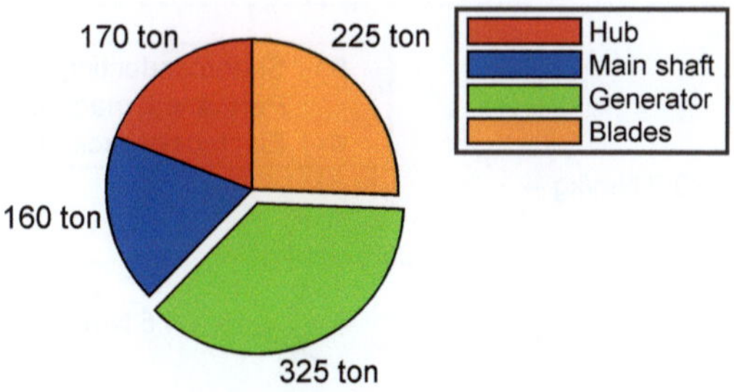

Fig. 1.16: Distribution of the tower head mass for a 10 MW permanent magnet wind turbine generator[80].

Data from different generators, including superconducting, permanent magnet and copper coil excited wind generators, in literature has been summarized and compared from weight, volume and efficiency. The results are listed in Table 1.7. It should be noted that many designs are from research papers, whereas some designs of the permanent magnet and copper coil excited generators are commercial products. However, it is believed that the collected data emphasizes the main difference of the three types of excitation generators. The weight and volume comparison is underlined in Fig. 1.17 and Fig. 1.18. The direct drive superconducting generators are lighter and smaller than the permanent magnet and copper coil excited generators, whereas the permanent magnet generators are lighter than the copper coil excited generators. The average torque to mass ratio of the superconducting generators is 64.5 Nm/kg, 29.8 Nm/kg for permanent magnet generators, and 20.2 Nm/kg for the copper coil excited generator. The average volume to torque ratio of the superconducting generator is 222.2 kNm/m^3, and 46.3 kNm/m^3 for the permanent magnet generator. In this case, the superconducting generator has the best potential to save weight and space.

The efficiency of different generators is compared in Fig. 1.19. In general, the efficiency of the superconducting generator is higher with higher torque output. In direct drive superconducting generators, the dominant losses are from armature winding, when copper is employed, whereas in the fully superconducting generator, the dominant losses are the refrigerator consumption, which is used to maintain the low temperature of the superconductors in both rotor and stator [88]. The main loss source for a permanent magnet generator is copper losses and permanent magnet losses caused by the harmonic contents in the airgap [97]. With no doubt the main losses of the copper coil excited generator is copper losses, which is more than 90% of the annual dissipation for a 3 MW generator when converter losses is not included [39].

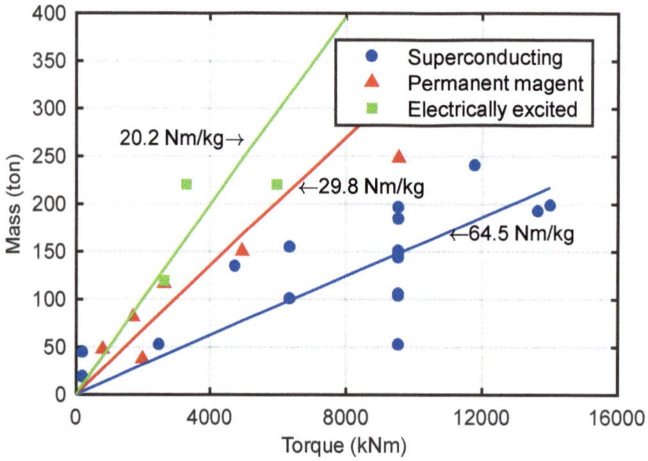

Fig. 1.17: Mass of different large direct-drive generators as a function of the torque from Table 1.7.

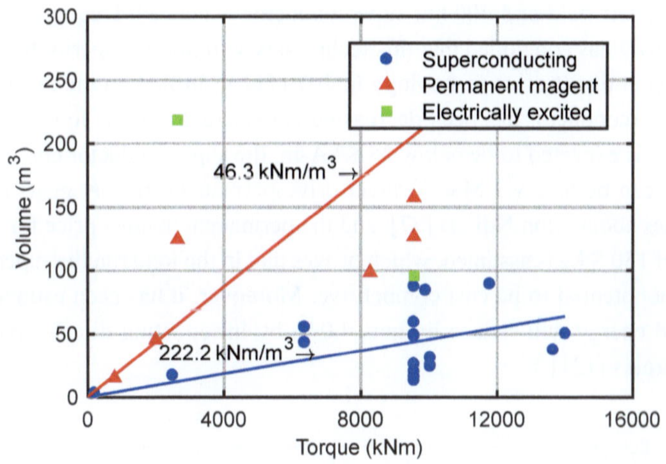

Fig. 1.18: Volume of different large direct-drive generators as a function of the torque from Table 1.7.

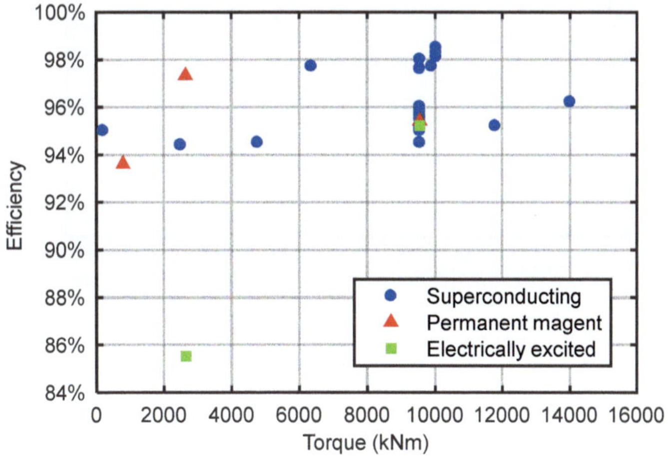

Fig. 1.19: Efficiency of different large direct-drive generators as a function of the torque from Table 1.7.

At present, the cost of the superconducting generator is expected to be higher than that of conventional generators. The total amount of superconductors in a generator with the same power rating is not identical, as illustrated in Fig. 1.20. It depends on the material type, quality of the superconductor, dimension of the superconductor, generator topology, design optimization objectives, and also working temperature. The superconducting length used in a 10 MW generator with YBCO is in the range of 10 to 800 km. Using the current price 100-1000 \$/kA-m of the YBCO as mentioned before, only the material price of the superconductor for a 10 MW generator is more than 4 M\$, if 100 A is assumed for the critical current at 77 K, self-field and 400 km superconductor is needed. Due to this reason, EU project INNWIND has concluded that this technology will not be superior to the expected permanent magnet direct drive technology in 2020 [120]. In fact, in the last ten years the price of the superconductor has been decreasing considerably. In the long-run, the superconducting tape is expected to be below 25 $/kA-m, the superconductor cost for a 10 MW generator then can be below 1 M$. A direct drive permanent magnet generator with the same rating uses about 8 ton NdFeB [97], and the permanent magnet price for this generator is 1.2 M$, if 150 $/kg is assumed, which proves that in the long run the superconducting generator has a potential to be cost competitive. Moreover, it has been estimated that the LCOE of wind energy will reduce by max.4.0 % by introducting direct-drive superconducting drive trains [121].

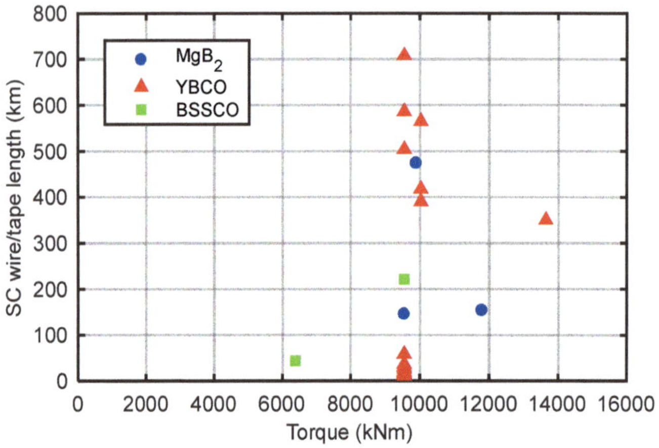

Fig. 1.20: Superconductor consumption as a function of torque from Table 1.7.

Table 1.7: Technical data of commercially available ReBCO

Affiliations	Power (MW)	Speed (1/min)	Mass (ton)	Volume (m3)	SC length (km)	Eff. (%)	Materials	Poles	Tempera-ture(K)	Bairgap (T)	Diameter (mm)	Topology
GE [61]	10	10	143	49	720	95-96	NbTi	36	4.3	2.6	4830	Syn.
Huazhong Univ. of Sci.&Tech.[14]	13.2	9	198	50	920	96.2	NbTi	40	4.2	2.1	7340	Syn.
DTU Wind Energy[98],[98]	10	9.65	52b	84	474	97.7	MgB2	32	10-15	1.5	5880	Syn.
Changwon National Univ.[100]	12	8.4	192	37	350	--	(RE)BCO	30	20	--	7800	Syn.
Kalsi Green power system[88]	10	10	52	20	146	98	MgB2	24	15/20	1.7	5000	Syn.
National Institute of AIST [101]	10	10	150	87[a]	26	97.6	(RE)BCO	48	40	1.0	10500	Syn.
Zenergy and Converteam [102],[62]	8	12	100	43	--	--	2G	--	30	--	5000	Syn.
AMSC [62],[63],[103]	10	10	150	<59	1.5 per poleset/36 km	96	YBCO	24	30	--	4500-5000	Syn.
AMSC [3]	10	11.54	312	98	--	--	PM	--	--	--	4300	Syn
Tecnalia[64],[104]	10	8.1	<240	89[a]	154	95.2	MgB2	48	20	1.5	10100[c]	Syn.
ECO 5	3.67	14	52	17	13500kAm	94.4	MgB2	32	20	1.74	5500	Syn.
Univ. of Edinburgh [103]	10	10	184	48	15/3.4	94.5	MgB2/YBCO	88	30	--	6630	Claw.
Delft Univ. of Tech.[105],[54]	10	10	325	126[a]	--	--	PM	160	--	--	5000	Syn.
Univ. of Edinburgh [106]	6	12	134	--	43	94.5	HTS	360	--	1.32	12000	Homo.
NewGen [107]	4	19	37	45	--	--	PM	--	--	--	9000[c]	Syn.
Univ. of Electron. Sci.&Tech. of China[108]	10	10	66[b]	13[a]	504	--	YBCO	24	20	4.5	5806	Syn.
Maki [109]	8	12	154	55	44	97.7	BSCCO	20	--	--	5000	Syn.
Changwon National Univ.[110]	10.5	10	105[b]	24	390	98.1	YBCO	24	20	--	5330	Syn.
Changwon National Univ. [110]	10.5	10	108[b]	25	418	98.3	YBCO	24	20	--	--	Syn.
Changwon National Univ. [110]	10.5	10	95[b]	31	565	98.5	YBCO	24	20	--	--	Syn.
Harakosan Z72[111]	1.5	18	47	15[a]	--	93.6	PM	60	--	--	4000	Syn.
Niigata Univ. [112]	10	10	50[b]	16[a]	572	--	HTS	8	20	2.5-4	3670	Syn.
Niigata Univ. [112]	10	10	55[b]	16[a]	856	--	HTS	12	20	2-3.5	3670	Syn.
Univ. of Tokoy [113]	10	10	--	26	--	--	HTS	12	20	2.2	4320	Syn.

Iran Univ. Sci.e and Tech.[92]	10	10	74 [b]	15 [a]	708	--	YBCO	8	20	3	3210	Syn.
Tokyo Univ. of Marine Sci. & Tech.[114]	10	10	103	48	10	95.1	YBCO	96	40	1.12	14100	Syn.
Tokyo Univ. of Marine Sci. & Tech. [114]	10	10	103	--	13	95.5	YBCO	96	50	--	--	Syn.
Tokyo Univ. of Marine Sci. & Tech. [114]	10	10	103	--	16	95.6	YBCO	96	55	--	-	Syn.
Tokyo Univ. of Marine Sci. & Tech. [114]	10	10	103	--	19	95.8	YBCO	96	60	--	--	Syn.
Tokyo Univ. of Marine Sci. & Tech. [114]	10	10	104	48	30	95.9	YBCO	96	68	1.12	14100	Syn.
Tokyo Univ. of Marine Sci. & Tech. [114]	10	10	105	48	58	96.0	YBCO	96	77	1.10	14100	Syn.
Tokyo Univ. of Marine Sci. & Tech. [114]	10	10	188	95.2	--	95.2	Copper	96	--	0.83	14100	Syn.
Hanyang Univ.[115]	5	230	18	3[a]	100	95.0	HTS	6	30	--	1747	Syn.
Hanyang Univ. [115]	5	230	44	7 [a]	50	95.0	HTS	6	30	--	2500	Homo.
Changwon National Univ.[116]	10	10	147	--	586	--	YBCO	24	20	--	--	Syn.
Changwon National Univ. [116]	10	10	196	--	222	--	BSCCO	24	20	--	--	Syn.
Changwon National Univ.[116]	10	10	304	--	--	--	PM	--	--	--	--	Syn.
Enercon [117]	4.5	13	220	--	--	--	Copper	--	--	--	--	Syn.
The Switch [117]	3.8	21	81	--	--	--	PM	--	--	--	--	Syn.
Georgia Institute ofTech. [117]	5	18	119	218 [a]	--	85.5	Copper	400	--	0.25	15310 [c]	Ind.
Georgia Institute of Tech. [117]	5	18	116	124 [a]	--	97.3	PM	400	--	0.7	11634 [c]	Syn.
KIT [97]	10	10	248	157	--	95.4	PM	200	--	0.8	9510	Syn.
GE Haliade[118]	6	11.5	150	--	--	--	PM	128	--	--	7600	Syn.
Enercon [119],[96]	7.58	12.1	220	--	--	--	Copper	--	--	--	--	Syn.

a The value is calculated based on the outer diameter and active stack length.
b The value is the active mass of the generator, which excludes structural mass.
c. The value is the armature diameter, instead of generator outer diameter

2 Design method

2.1 Design equations

In the design process of the superconducting generator, the terminal voltage, speed, and power are usually determined by the power system as design requirements. The main task of the design is to find out the most optimal machine within the design constraints, including economic and reliability considerations. As a start, the relationship of the output power with the generator size, electrical loading, magnetic loading and speed is expressed by eq (2.1) [33], which can be also referred as sizing equation.

$$\frac{P_N}{\eta} = \frac{\pi}{2} \alpha_{DC} A_a B_{\delta max} D_a^2 l_{ef} \Omega_N \tag{2.1}$$

where, D_a is the outer diameter of rotor; l_{ef} is the effective length of stator stack lamination; Ω_N is the rated mechanical angular speed, which is equal to $2\pi n_N/60$, n_N is the rated rotational speed in 1/min; P_N is the output power; η is the assumed efficiency of the generator; $B_{\delta max}$ is the magnitude of the radial flux density in the middle of air gap, which is generated by the superconducting field coil in the rotor and a value is set first based on experience and literature; A_a is the electrical loading of armature winding, which is chosen by the cooling capacity of the stator; and α_{DC} is the relative pole width for DC.

The rated current I_{aN} of the generator is calculated as eq. (2.2).

$$I_{aN} = \frac{P_N}{U_N} \tag{2.2}$$

where, U_N is the rated voltage.

Generally, the following design step is to determine the dimensions of the machine. The rational approach is to set the ratios of effective length, l_{ef}, to the outer diameter of the rotor, D_a. After that, the dimensions, such as the rotor dimension, stator dimension and the armature winding design, can be selected. The dimension parameters needed in the design of the superconducting generator are described in Fig. 2.1, which displays a transacted cross section of the round generator with one pole pair as an illustration. The gap between the coilformer and the dewar is the space for vacuum which reduces the conduction heat load to the superconducting coils. Normally, the pole and teeth are made of non-magnetic materials to support the copper armature winding and superconducting coils in the superconducting generator, for example, the teeth are generally made of glass fibre and the poles are made of G10.

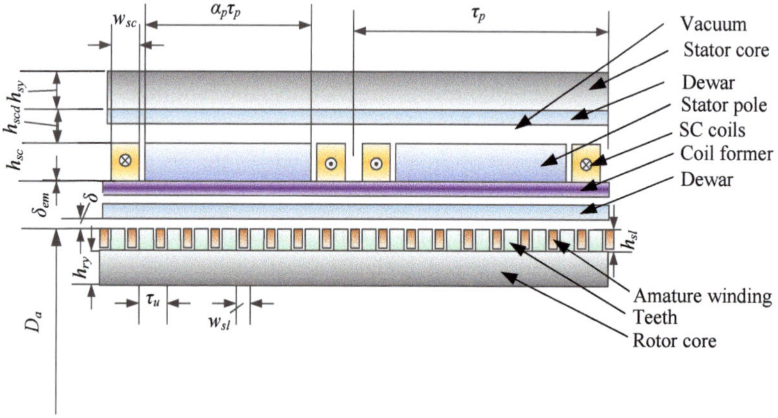

Fig. 2.1: Dimension parameters in a superconducting DC generator.

The dimension of the slot can be derived by eq. (2.3) and (2.4).

$$w_{sl}h_{sl} = \frac{I_{slot}}{j_a\alpha_u} = \frac{2uN_a I_{aN}}{2aj_a\alpha_u} \tag{2.3}$$

$$N_a = \frac{N_{cond}}{2uZ_u} = \frac{\pi D_a A_a}{(I_{aN}/2a)(2uZ_u)} = \frac{2a\pi D_a A_a}{I_{aN}(2uZ_u)} \tag{2.4}$$

where, w_{sl} is the width of slot; h_{sl} is the height of slot; I_{slot} is the current in a slot; j_a is the current density in armature winding; α_u is the slot filling factor; u is the number of coils per slot; a is the number of branch pairs; N_a is the number of turns per armature coil; I_{aN} is the rated current; N_{cond} is the number of conductors in series; Z_u is the number of slots, which is selected based on the material strength of teeth and the performance of the commutation.

When iron cores are employed in stator and rotor, it is essential to select the height of the stator and rotor core in radial direction. Since the iron stator and rotor core confine the flux in the machine, the flux linkage per pole φ_m is calculated first as eq. (2.5).

$$\varphi_m = B_\delta \tau_p l_{ef} \tag{2.5}$$

where B_δ is the average radial flux density in the middle of air gap; and τ_p is the pole pitch which can be calculated as eq. (2.6).

$$\tau_p = \frac{\pi\left(D_a + 2\delta_{em} + h_{sc}\right)}{2p} \tag{2.6}$$

where, δ_{em} is the effective length of air gap, including the thickness of the cryostat wall, and the physical air gap; p is the number of pole pairs; and h_{sc} is the height of one side of superconducting field coil.

Then the height of stator core and rotor core can be calculated as eq. (2.7) and (2.8), if they are made of ferromagnetic materials.

$$h_{sy} = \frac{\varphi_m}{2k_{Fe} l_{ef} B_{sy}} \tag{2.7}$$

$$h_{ry} = \frac{\varphi_m}{2k_{Fe} l_{ef} B_{ry}} \tag{2.8}$$

where, h_{ry} is the height of rotor yoke; h_{sy} is the height of stator yoke B_{sy} is the magnetic flux density in the stator yoke; B_{ry} is the magnetic flux density in the rotor yoke; and k_{Fe} is the space factor of the core, generally it is 0.98[33].

The stator generates the main flux density in the machine. When the rotor coils rotate, the armature winding in the rotor moves through the magnetic flux, and a voltage, the back electromotive force (back-EMF), is induced in the armature winding. According to the law of induction the back-EMF V_{aN} can be calculated as eq. (2.9).

$$V_{aN} = \sum_{n_c=1}^{N_{cond}} B_{\delta n}\left(x\right) l_{ef} v \tag{2.9}$$

where, n_c is an integer, which is 1, 2, …, N_{cond}; $B_{\delta n}(x)$ is the airgap radial flux of n_c-conductor in x position, as show in Fig. 2.2, in which the brush is located in the geometrical neutral line between the north and south pole; and v is the velocity the conductor cuts through the magnetic field, which can be calculated as eq. (2.10).

To make the expression of the back-EMF easy to follow, the average value, instead of the circumferential distribution of the flux density per pole, is preferred to use. The average value of flux density B_δ can be derived as in eq. (2.11).

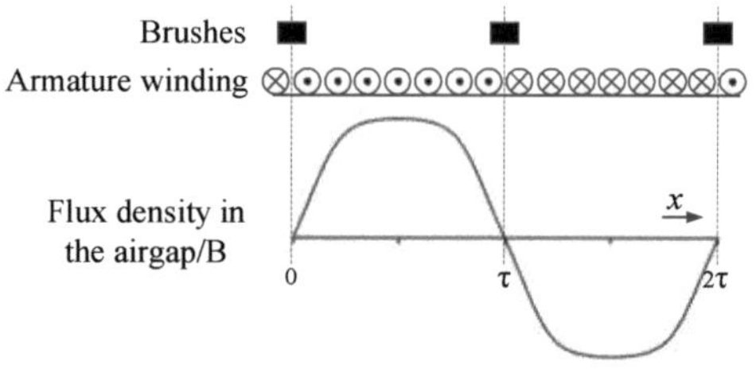

Fig. 2.2: Distribution of airgap flux density and conductors.

$$v = \frac{2p\tau_p n_N}{60} \tag{2.10}$$

$$B_\delta = \frac{1}{N_{cond}} \sum_{n_c=1}^{N_{cond}} B_{\delta n}(x) = \frac{2a}{2uZ_u N_a} \sum_{n_c=1}^{N_{cond}} B_{\delta n}(x) \tag{2.11}$$

If the armature winding is connected to the load, current will flow in it due to the back-EMF. As a result of Ampere's law, the armature winding generates a magnetic field. Due to the force between the magnetic field of the rotor and stator, there is a tendency to align the two magnetic fields. Hence, torque and power is produced. In the superconducting DC generator, the pole and teeth are usually made of nonmagnetic materials as fermentioned, which enlarge the effective air gap. Moreover, the cryostat to house the superconducting coils makes the effective air gap even larger. Consequently, different from conventional DC generator, the influence of armature reaction on the no-load flux density is negligible. Hence, the generated power P_g without losses is calculated as eq. (2.12), based on the.no-load flux density.

$$P_g = V_{aN} I_{aN} \tag{2.12}$$

By combining eq. (2.5) and (2.9) with eq. (2.12), the calculated power P_g is

$$P_g = \frac{2puZ_u n_N I_{aN} N_a \varphi_m}{60a} \tag{2.13}$$

In order to calculate the efficiency of the generator, copper losses, iron losses, and additional losses also need to be calculated. In this part, the losses of the cryostat walls and AC losses in the superconducting coils are neglected as the effective air gap is large and the harmonics to cause eddy current loss in the cryostat walls and superconducting coils decay rapidly with distance. To calculate the copper losses, the length of the armature coil per turn is estimated. The shape of an armature coil is shown in Fig. 2.3

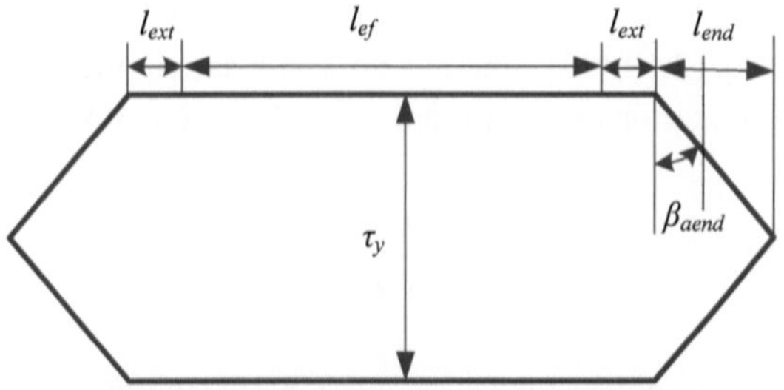

Fig. 2.3: Sketch and main dimensions of an armature coil.

The axial length of the end armature winding per turn in one side lend is calculated as

$$l_{end} = \frac{\tau_y}{2}\tan\beta_{aend} = \frac{\sqrt{3}}{6}\left[\frac{y_1}{u}\tau_u + \left(y_1 - \frac{y_1}{u}\right)w_{sl}\right] \tag{2.14}$$

where, τ_y is the distance of the coil span; β_{aend}, as shown in Fig. 2.3, is the bending angle of the armature end winding to the coil span, which is assumed to be 30 degree; y_1 is the coil span as the number of coil sides.

It can be calculated from Fig. 2.3 that the length of the copper winding per turn l_{aw} is

$$l_{aw} = 4l_{end} / \sin(\beta_{aend}) + 2l_{ef} + 4l_{ext} = 8l_{end} + 2l_{ef} + 4l_{ext} \tag{2.15}$$

where l_{ext} is the extension length of armature winding per coil in direction part.

Then the resistance of an armature coil per turn R_{at} is expressed as eq. (2.16).

$$R_{at} = \rho_{20}\left[1+(T_a - 20)C_a\right]\frac{l_{aw}}{I_{aN}/(2aj_a)} \tag{2.16}$$

where, ρ_{20} is the resistivity of copper in 20 Celsius, which is 1.724E-08 Ωm; T_a is the working temperature of armature winding; C_a is the temperature coefficient of resistance, which is 0.00393/ Celsius.

Hence, the resistance of the copper winding R_a can be expressed as eq. (2.17). Copper losses P_{cu} can be resulted from eq. (2.14)-(2.17), shown as eq. (2.18).

$$R_a = R_{at} \times \frac{N_a u Z_u}{2a} \times \frac{1}{2a} = \rho_{20}\left[1+(T_a - 20)C_a\right]\frac{l_{aw} N_a u Z_u j_a}{2a I_{aN}} \tag{2.17}$$

$$P_{cu} = I_{aN}^2 R_a \tag{2.18}$$

If rotor core and stator core are made of ferromagnetic materials, the iron losses P_{Fe} can be roughly estimated by eq. (2.19)[33]. There is no iron losses in the stator core as the fund-mental magnetic field is stationary if harmonics are not taken into consideration.

$$P_{Fe} = k_{ry} P_{15} \left(\frac{B_{ry}}{1.5} \right)^2 m_{ry} f^{1.3} \tag{2.19}$$

where, k_{ry}, is correction coefficient for P_{Fe} calculation, which are 1.8; B_{ry}, is the magnitude of flux density in rotor yoke; m_{ry} is the active material weight of rotor yoke, P_{15} is the losses of ferromagnetic materials per mass unit at 1.5T, 50 Hz; f is the frequency of the generator.

The additional losses P_{add} can be calculated by empirical equation, shown as eq. (2.20)[33].

$$P_{add} = 0.01 \times P_N \tag{2.20}$$

The brush losses P_b can be calculated by as (2.21)[33].

$$P_b = 2U_b I_{aN} \tag{2.21}$$

Where, U_b is the voltage of brushes.

If the friction and windage loss P_{fw} is neglected, the efficiency of the generator η_g can be calculated as eq. (2.22).

$$\eta_g = 1 - \frac{P_{cu} + P_{Fe} + P_{fw} + P_b + P_{add}}{P_g} \tag{2.22}$$

The sketch and main dimensions of a superconducting coil are shown as in Fig. 2.4. From Fig. 2.4 the length of the superconducting coil per turn can be calculated as eq. (2.23).The supercondcutingcoil is ramped up and then continously running with a voltage source to avoid unrealistic induced voltages and demagnetisation in coils.

Then the length of the superconducting tape L_{sc} is derived as eq. (2.24).

Another key factor needs to be considered in the generator design is the weight of the generator. The materials weight of the generator m_g is derived as eq. (2.25).

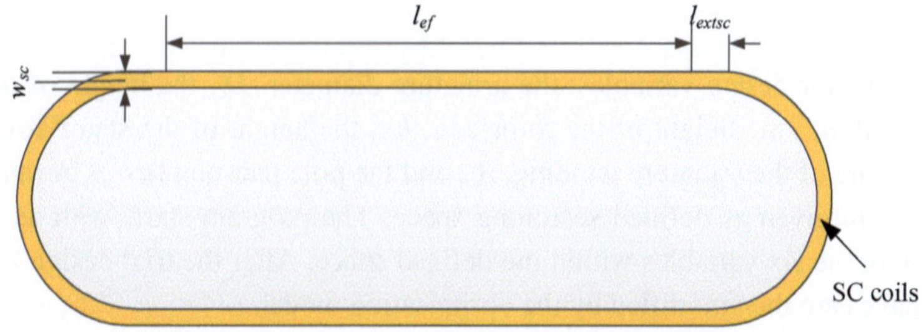

Fig. 2.4: Sketch and main dimensions of a superconducting coil.

$$L_{scw} = 2l_{ef} + 4l_{extsc} + \pi\left(\alpha_p \tau_p + w_{sc}\right) \tag{2.23}$$

where L_{scw} is the length of the superconducting coil per turn; l_{extsc} is the extension length of a superconducting coil in direction part; and w_{sc} is the width of one side of superconducting coil, α_p is the pole arc coefficient.

$$L_{sc} = 2pN_{sc}L_{scw} = 2pN_{sc}\left[2l_{ef} + 4l_{extsc} + \pi\left(\alpha_p \tau_p + w_{sc}\right)\right] \tag{2.24}$$

where, N_{sc} is the number of turns per superconducting coil.

$$m_g = m_{sy} + m_{ry} + m_{Cu} + m_{sc} + m_{frame} \tag{2.25}$$

where m_{Cu} is the active material weight of copper winding; m_{sc} is the active material weight of superconducting winding; m_{sy},and m_{ry} are the active material weight of stator yoke and rotor yoke, respectively; and m_{frame} is the structural material weight of the generator frame.

2.2 Design process

The fundamental design process is presented as Fig. 2.5. This design program uses iterative calculation to adjust the superconducting coil current to meet the required superconducting tape characteristics. Then the program is run to design the active part of the generator by adjusting effective axial length. Thereafter the generator performance is calculated, including efficiency, volume, weight, superconducting weight length and so on. Moreover, to accurately calculate the weight of the generator, a mechanical model is employed to determine the thickness of the frame to safely support the stresses on the stator and rotor components. Finally a trial design based on the input parameters is saved for the optimisation. The explanations of the design process by step are as follow.

Step 1:

There are six major design variables: the armature diameter, D_a, the height of the superconducting coil, h_{sc}, the height of the rotor coil, h_{ry}, the height of the stator coil, h_{sy}, the electrical loading of the armature winding, A_a, and the pole pair number, p. A range of the six variables are given as defined searching space. The program starts with an arbitrary combination of the six variables within the defined space. After the trial design is located, the six variables are then modified by the optimisation model.

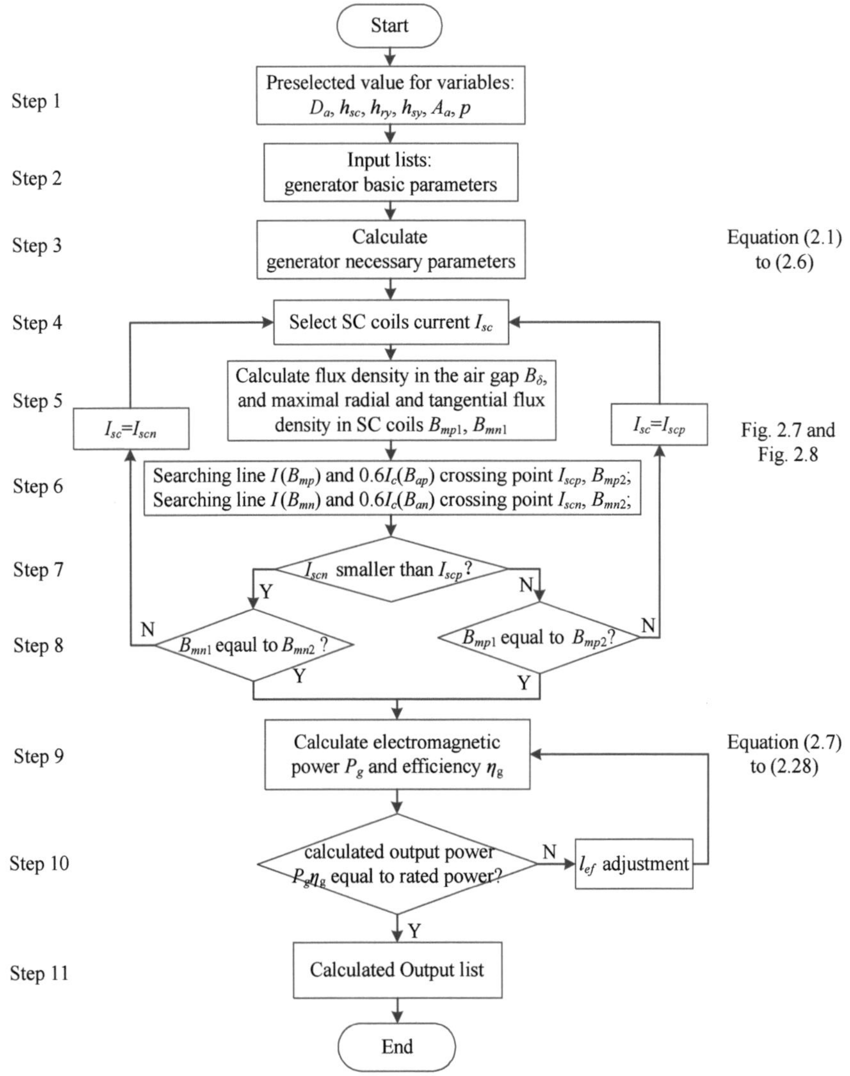

Fig. 2.5: Flowchart of design process.

Step 2:

The predefined parameters about generator size, power, superconducting coils, armature winding and stator and rotor core are input also at the beginning of the program. The input parameters are summarized as illustrated from Table 2.1 to Table 2.5.

Step 3:

Through equation (2.1) to (2.8), more parameters which are needed in the machine design are calculated, such as slot size, copper winding design etc. The parameters calculated in this step are listed in Table 2.6.

Table 2.1: Parameters of superconducting coils

Symbol	Explanation	Unit
α_p	Pole arc coefficient	--
α_{f1}	Safety margin for SC coils	--
α_{f2}	Packing factor for SC coils	--
d_{esc}	Mass density of SC wires	kg/m^3
I_{sc}	Operating current of SC coils	A
$I_c(B_{an}, T_{sc})$	Critical current and applied flux density curve (H//ab)	A
$I_c(B_{ap}, T_{sc})$	Critical current and applied flux density curve (H//c)	A
l_{extsc}	Extension length of a SC coil in direction part	m
S_{sc}	Cross section area of SC tape	m^2
T_{sc}	Working temperature of SC coils	K
w_{sc}	Width of one side of a SC coil	m

Table 2.2: Parameters used in sizing equation and losses

Symbol	Explanation	Unit
P_N	Output power	W
U_N	Rated voltage	V
n_N	Rated rotation speed	1/min

Table 2.3: Parameters of the air gap

Symbol	Explanation	Unit
δ	Physical air-gap length	m
δ_{em}	Effective air-gap length	m
h_{scd}	Height of superconducting coil to stator core	m

Table 2.4: Parameters of armature winding

Symbol	Explanation	Unit
a	Number of branch pairs	--
α_u	Slot filling factor	--
d_{ec}	Copper mass density	kg/m^3
j_a	Current density in armature winding	A/m^2
l_{ext}	Extension length per turn in direction part	m
T_a	Working temperature of armature winding	Celsius
u	Coils per layer in a slot	--
w_{sl}/τ_u	Width of slot to slot pitch	--
q	Slot number per pole	--

Table 2.5: Parameters of generator cores

Symbol	Explanation	Unit
d_{ey}	Mass density of silicon lamination	kg/m^3
k_{Fe}	Correction coefficient for P_{Fe}	--
k_{ry}	Correction coefficient for P_{Fe}	--
k_{sy}	Correction coefficient for P_{Fe}	--
P_{15}	Losses of iron per mass unit at 1.5T, 50 Hz,	W/kg
μ_0	Permeability of air	H/m
μ_{rr}	Relative permeability of rotor core	--
μ_{rs}	Relative permeability of stator core	--

Table 2.6: Calculated parameters through equation (2.1) to (2.6)

Symbol	Explanation	Unit
l_{ef}	Effective length of stator stack lamination	m
h_{sl}	Height of slot	m
w_{sl}	Width of slot	m
u	Number of coils in a slot	--
Z_u	Number of slots in rotor	--
N_a	Number of turns per coil	--
I_{aN}	Rated armature current	A
τ_p	Pole pitch	m
N_{sc}	Number of turns per superconducting coil	--

Step 4:

The current of the superconducting coils I_{sc} is preselected first. Then the number of turns N_{sc}, and flux density in the air gap B_δ can be calculated according to equations of the electromagnetic model in Chapter 3. The current of the superconducting coils I_{sc} is modified to an appropriate value for a certain size of the superconducting coils based on the performance of the superconducting tapes, that is the $I_c(B_a,T_{sc})$ curves.

Step 5:

The process to calculate flux density in the generator is shown in Fig. 2.6. According to the materials used in stator and rotor core, different equations from Chapter 3 are employed. Moreover, in the iron core generator, the working relative permeability of the rotor and stator core, u_{rr} and u_{rs}, are found by iterative calculations. In this step, The *B-H* curve of the ferromagnetic material is employed, as shown in Fig. 2.7.

Step 6:

In order to find the operating current of the superconducting coil, the superconducting coil load line, which is the superconducting coil current against magnetic flux density, and the tape performance, that is, critical current against applied flux density of the tape, are drawn in the same figure. In Fig. 2.8, Line $Ic(B, 30K)$ is the critical current of the superconducting tapes versus the radial flux density ($H//$ab), while Line $0.6Ic(B, 30K)$ is the critical current with safety margin of 60% versus the radial flux density ($H//$ab). The load line of the superconducting coil, Line $I(B_{mp})$ ($H//$ab), is also plotted in the same figure. The crossing point of the load line and the 60% of critical current line, I_{scp}, is the operating current for the superconducting coil in terms of the radial flux density. As the superconducting tapes in the generator is also subjected to the tangential flux density ($H//$c) besides radial flux density ($H//$ab), then the operating current I_{scn} in terms of the tangential flux density also needs to be found. The process to find out the operating current I_{scn} is similar to that of the operating current I_{scp} , as shown in Fig. 2.9.

Step 7:

As a safety margin of 60% for the superconducting coil is selected in terms of both radial and tangential flux density, the smaller value of the operating current in terms of the radial and tangential flux density, I_{scp} and I_{scn}, is the true operating current for the coil.

Step 8:

To find the load line of a superconducting coil needs many calculations with different superconducting coil currents. To simplify the calculation and save time, a linear load line is assumed at first and will be iterated to the real value. Hence, the load line can be drawn by a given value of the superconducting coil current I_{sc} and its applied radial and tangential flux density, B_{mp1} or B_{mn1}. Then the crossing point (I_{scp}, B_{mp2}) or (I_{scn}, B_{mn2}) from the load line and the 60% critical current line is calculated. However, whether the crossing point is in the real load line, instead of the assumed linear load line, is checked by comparing B_{mp1} & B_{mp2} or B_{mn1} & B_{mn2}. If they are not equal, the current of the superconducting coil I_{sc} has to be modified. Otherwise, the program can move on.

Step 9:

Through equation eq. (2.9) to (2.22), the electromagnetic power P_g and losses of the generator, including copper losses P_{cu}, additional losses P_{add}, iron lossesP_{Fe} and brush losses P_b can be calculated. Then the efficiency η_g and output power are also obtained.

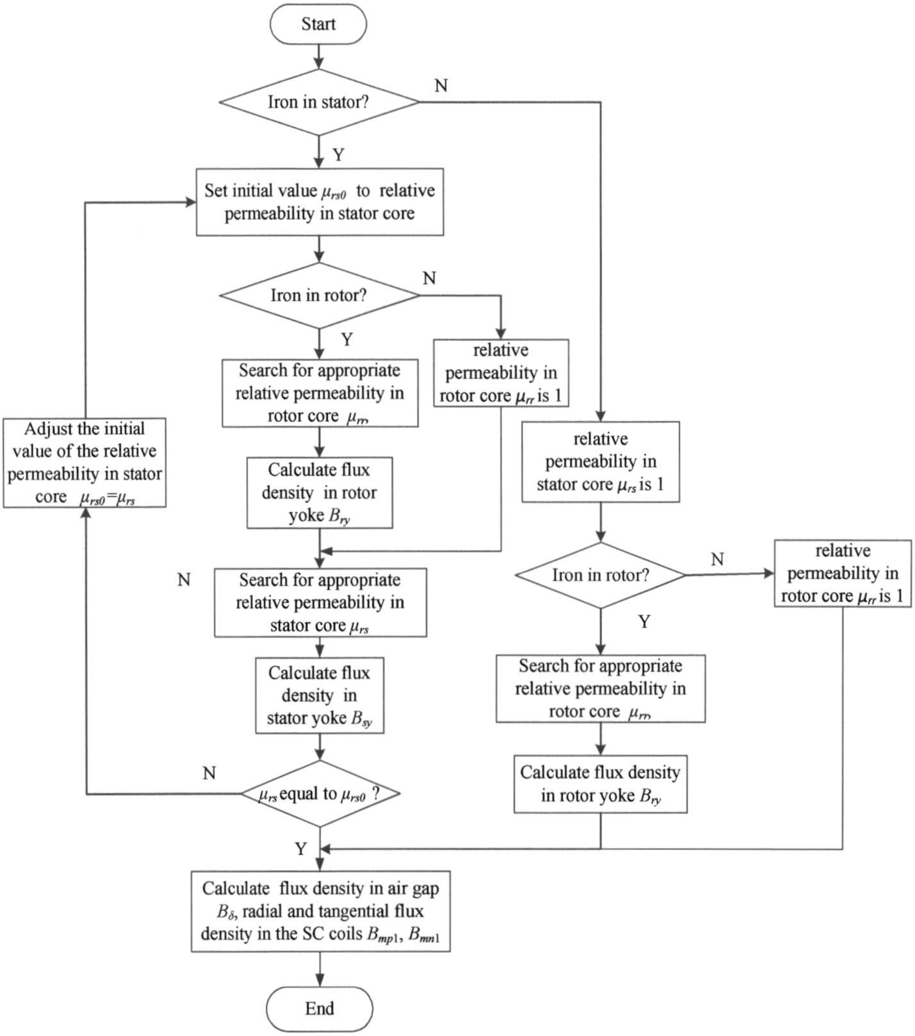

Fig. 2.6: Flowchart to calculate flux density in the generator.

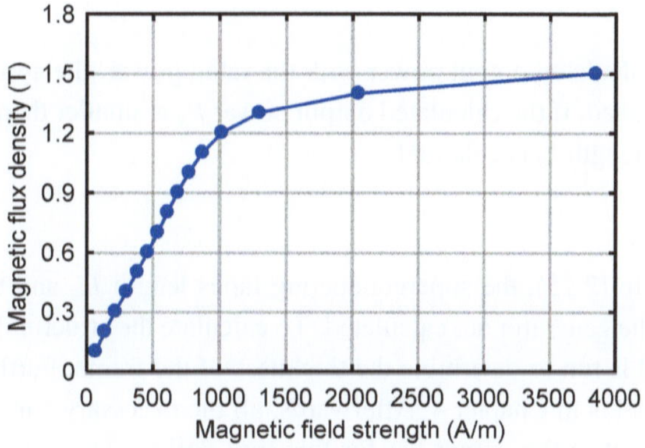

Fig. 2.7: *B-H* curve of M235-35 @50 Hz.

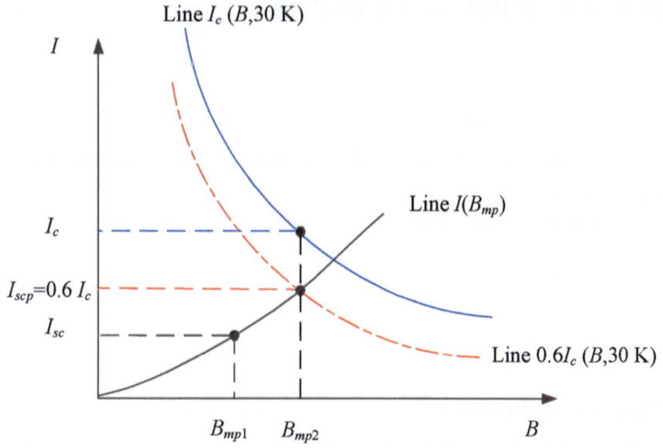

Fig. 2.8: Operating current of a superconducting coil @ H//ab.

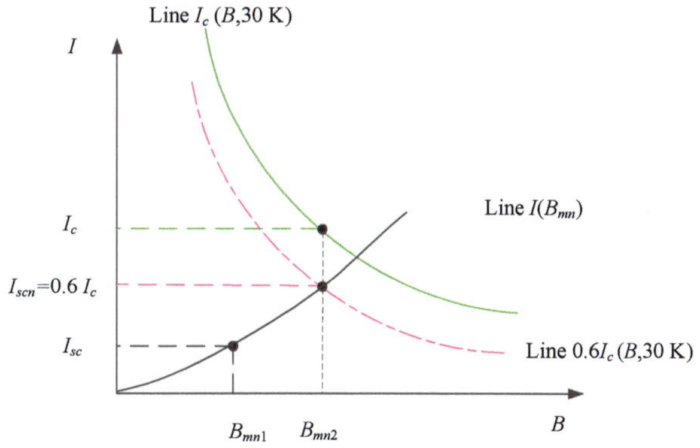

Fig. 2.9: Operating current of a superconducting coil @ H//c.

Step 10:

By comparing the calculated output power with the rated power, the appropriate effective axial length l_{ef} is reached. If the calculated output power P_g is smaller than the target output power P_N, the axial length l_{ef} is enlarged.

Step 11:

By using eq (2.23) to (2.25), the superconducting tapes length L_{sc} and the weight of the active materials in the generator are calculated. To calculate the structural material weight, a mechanical model is run to determine the thickness of the frame. Further description of the mechanical model is in Chapter 4. Afterwards, all the necessary parameter of the generator design is output in the output list for this trial design. The output parameters are summarized from Table 2.7 to 2.10. The trial design, represented by the six prime variables, is then evaluated by the optimization method according to a design objective function in Section 2.3 to decide how to adjust the six variables.

Table 2.7: Parameters of losses

Symbol	Explanation	Unit
P_{Fe}	Iron losses	W
P_b	Brushes losses	W
P_{add}	Additional losses	W
P_{fw}	Friction and windage loss	W
P_{cu}	Copper losses	W
P_g	Electromagnetic power	W
η_g	Calculated generator efficiency	--

Table 2.8: Parameters of dimensions

Symbol	Explanation	Unit
D_a	Outer diameter of the rotor	m
l_{ef}	Effective length of stator stack lamination	m
h_{sl}	Height of slot	m
w_{sl}	Width of slot	m
h_{sc}	Height of the superconducting coil	m
w_{sc}	Width of superconducting coil	m
h_{ry}	Height of rotor core	m
h_{sy}	Height of Stator core	m
δ	Physical air-gap length	m
δ_{em}	Effective air-gap length	m
h_{scd}	Height of superconducting coil to stator core	m

Table 2.9: Generator design parameters

Symbol	Explanation	Unit
B_δ	Average value of radial flux density in air gap	T
A_a	Electrical loading of armature winding	A/m
a	Number of branch pairs	--
α_u	Slot filling factor	--
u	Number of coils in a slot	--
Z_u	Number of slots in rotor	--
N_a	Number of turns per coil	--
I_{aN}	Rated armature current	A
R_a	Resistance of armature winding	ohm
α_p	pole arc coefficient	--
I_{sc}	Operating current of SC coils	A
N_{sc}	Number of turns per superconducting coil	--
B_{mp}	Maximum flux density in SC field coil in direction of H//ab	T
B_{mn}	Maximum flux density in SC field coil in direction of H//c	T
L_{sc}	Length of superconducting field coils	m
p	Pole pair number	--

Table 2.10: Parameters of weights

Symbol	Explanation	Unit
m_{sy}	Active material weight of stator core	kg
m_{ry}	Active material weight of rotor core	kg
m_{Cu}	Active material weight of copper winding	kg
m_{sc}	Active material weight of superconducting coils	kg
m_g	Material weight of the generator	kg
m_{frame}	Structural material weight of the generator frame	kg

2.3 Optimization method

2.3.1 Introduction of the optimization method

From the analysis above it can be seen that many geometric dimensions and material characteristics influence the performance of a superconducting DC generator. In the design process of the generator, a large input data for the variables is necessary. Generally, a specific design is finished by the combination of the analytical equations and finite elements analysis software. The different input parameters, which are discrete and continuous variables of a multi-dimensional space, make the optimum design a difficult and challenging task. By changing only one parameter at one time, a data base of different designs for the same generator is established. The optimal design is selected based on the user's requirements, which are related to the cost, weight, volume, torque of the generator and so on. These make the design more complicated as the optimisation is a nonlinear and multi-objectives minimisation problem. Hence, the design process is very time consuming, especially if the finite element method is involved to calculate the flux distribution and performance of the generator.

With the advent of the large-scale multi-processing computers with high computing efficiency, evolutionary and intelligence-based optimisation methods, which provide optima for multi-dimensional and nonlinear functions, are proposed, for example, particle swarm optimisation (PSO). Although these methods are widely used in different fields, such as medical science and engineering design, the application in electric machines, especially superconducting generators is just beginning. In [122] PSO was used to minimize the cogging torque without deterioration the machine performance for a transverse flux permanent magnet (PM) machine. Further, PSO was proposed to determine optimum widths and displacements of PM pieces to minimize cogging torque without noticeable loss in the output [123]. A PM type transverse flux linear motor was optimized in terms of weight, thrust force and detent force [124]. In [125] the PSO was integrated with the electromagnetic and thermo-mechanical design for surface mount PM machines. Moreover, the PSO was employed to axially laminated anisotropic rotor synchronous reluctance motor for traction applications in [126].They have all proved that the PSO is effective to solve electrical machine optimization problem. Besides, PSO was compared with the genetic algorithm for PM synchronous machines [127],[128]. It shows that PSO is better at computational efficiency and solution optimality [123], [127].Therefore, in this section, the evolutionary optimisation algorithm PSO is adopted in the design of the superconducting DC generator to find the most suitable design.

PSO was put forward by Kennedy and Eberhart in 1995 as a population-based stochastic search and optimisation methods, which originated from the computer simulating the social and individual activities of birds or fish [129],[130]. In PSO, the system, which has the collective behaviours and social characteristics, is called "swarm" and the moving member in the system is denoted "particle". In the searching space, every position of the particle has a value for the objective function. Each particle in the swarm has a random direction to travel through the searching space and remember its best position so far achieved. The particles of the swarm can interact with each other and adjust their positions according to the good positions of their neighbours. In the end, all the particles tend to cluster together in a better position in the searching space, which has the optima value of the objective function. The optimisation procedure is described as shown in Fig. 2.10.

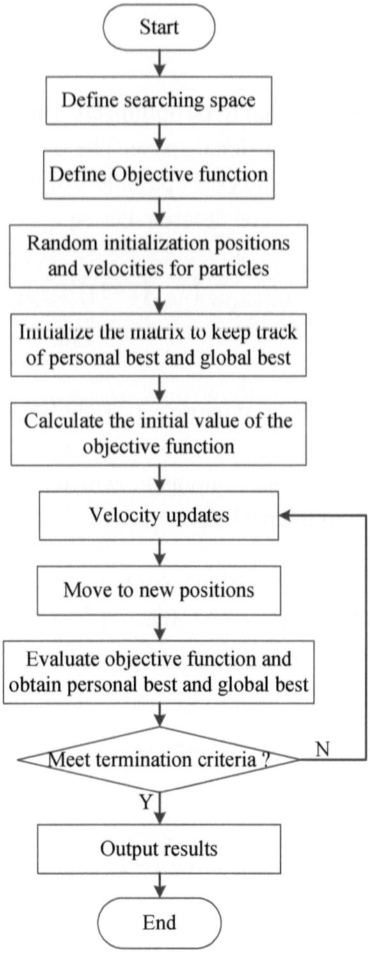

Fig. 2.10: PSO optimization procedure.

Firstly, the multi-dimensional searching space is defined, which are the paramount variables selected for the generator design and their scopes of domain. Secondly, the objective function is defined, which consists of the torque per length of the superconducting tapes, the torque per weight, the torque per volume and efficiency. Thirdly, the random swarm positions and velocities are initialized. Fourthly, the matrix to keep track of the best position of each particle throughout history and the best global solution is also initialized. Fifthly, the initial value of the objective function of each particle is calculated. Sixthly, the particles velocities are updated according to their best position in history and the global best solution. Seventhly, the new positions of the particles are calculated and the particles are moved to the new position. Eighthly, the objective function is calculated and evaluated by comparing the global best solution and personal best position. Then step six is repeated until the termination criteria are satisfied.

Let $f_{objective}$ denotes the objective function to be optimized. The objective function is to find the maximum value in the multi-dimensional space. The particle number in the swarm is represented by *n_particle*. The personal best of particle k of jth dimensional component in the iteration step i+1, $pbest_{k,j}(i+1)$, can be described as eq. (2.26).

$$pbest_{k,j}(i+1) = \begin{cases} pbest_{k,j}(i) & if\ f_{objective}\left(x_{k,j}(i+1)\right) < f_{objective}\left(pbest_{k,j}(i)\right) \\ x_{k,j}(i+1) & esle \end{cases} \tag{2.26}$$

where $pbest_{k,j}(i+1)$ is the personal best of particle k of jth dimensional component j in the iteration step i+1; k is the kth particle in the swarm, which is 1, 2, ..., *n_particle* ; i is the ith iteration step; j is the jth dimensional component, which is 1,2,..., *n_dimension*; $x_{k,j}(i)$ is the jth dimensional component of the position of particle k at the iteration step i.

The global best of the swarm of jth dimensional component in the iteration step i, $gbest_j(i)$, is a value among $\{pbest_{1,j}(i), pbest_{2,j}(i), pbest_{3,j}(i),\ldots, pbest_{n\_particle,j}(i)\}$,which makes $f_{objective}(gbest_j(i))$ the maximum value among $\{f_{objective}(pbest_{1,j}(i)), f_{objective}(pbest_{2,j}(i)), pbest_{3,j}(i),\ldots, f_{objective}(pbest_{n\_particle,j}(i))\}$.

The velocity and position updates, $v_{k,j}(i+1)$ and $x_{k,j}(i+1)$, of each particle in the searching procedure are expressed by eq. (2.27) and (2.28).

$$\begin{aligned} v_{k,j}(i+1) = {} & w(i)v_{k,j}(i) + c_1 r_1\left(pbest_{k,j}(i) - x_{k,j}(i)\right) \\ & + c_2 r_2\left(gbest_j(i) - x_{k,j}(i)\right) \end{aligned} \tag{2.27}$$

$$x_{k,j}(i+1) = x_{k,j}(i) + v_{k,j}(i+1) \tag{2.28}$$

where w is the inertial weight; c_1 and c_2 are the acceleration constants; r_1 and r_2 are random variables with a uniform distribution (0,1). In the velocity updates, the first term is the momentum component representing the contribution of previous velocity, the second term is the cognitive component representing the particle's own experience, and the third term is the social component representing the guidance by particle with global best value [129].

2.3.2 Design variables

The geometric dimensions of the generator are shown as Fig. 2.1. To implement the PSO in the design of the generator, variables selected are the height of stator core h_{sy}, the height of rotor core h_{ry}, the outer diameter of rotor D_a, the height of the superconducting coil h_{sc}, and the electrical loading of armature winding A_a. When the rotor or stator core is made of non-magnetic materials, the height of stator core h_{sy}, and the height of rotor core h_{ry} are not variables in the optimisation design. The ranges of the parameters for a 10 kW generator are listed in Table 2.11 as an example.

Table 2.11: Design control parameters

Symbol	Explanation	Range
D_a	Outer diameter of rotor	100-250 mm
h_{sc}	Height of the superconducting coil	4-40 mm
h_{ry}	Height of rotor core	10-60 mm
h_{sy}	Height of stator core	10-60 mm
A_a	Electrical loading of the armature winding	25-65 kA/m [33]
p	Pole pair number	2-6

More constraints for the design of the generator are needed. For example, the maximum flux density is below 1.8 T when the stator and rotor core are made of ferromagnetic materials. In order to have enough space to install the superconducting coils in the cryostat as the superconducting coils are manufactured into racetrack shape, the inner radius of the superconducting coil r_{sci} is limited by eq. (2.29) and the inner radius of the stator iron yoke r_{si} is limited by eq. (2.30).

$$r_{sci} = \left(\frac{D_a}{2} + \delta_{em}\right) / \cos\left(\frac{90°}{p}\right) \tag{2.29}$$

$$r_{si} = r_{sci} + h_{sc} + h_{scd} \tag{2.30}$$

Where, r_{sci} is the inner radius of the superconducting coil, D_a is the outer diameter of rotor, δ_{em} is the effective air-gap length, p is the number of pole pairs, r_{si} is the inner radius of the stator iron yoke, h_{sc} is the height of the superconducting coil, and h_{scd} is the height of superconducting coil to stator core.

2.3.3 Design objective function

To design the superconducting DC generator, the aims are maximum output power, least superconducting tapes usage, lightest weight, smallest volume, and highest efficiency. However, when one objective is optimal, for example, the usage of the superconducting tapes is least, volume, weight and efficiency are not in the optimal value. In this case, the balance between the four objectives is needed. In order to arrange the four objectives in the same function to have a maximum value, they are modified into four indexes, which are the length of the superconducting tapes, the mass, the volume, and efficiency. Each index is first normalized to the range between 0 and 1 within their constraints. The objective function is expressed as eq. (2.31), and the optimisation is to find its maximum value within the ranges of variables.

$$f_{objective} = w_{tl} f_{tl} + w_{tm} f_{tm} + w_{tv} f_{tv} - w_e f_e \tag{2.31}$$

where, $f_{objective}$ is the objective function to be optimized; f_{tl} is the normalized value of the superconducting tapes length within the constraints; f_{tm} is the normalized value of the machine mass within the constraints; f_{tv} is the normalized value of the machine volume within the constraints; f_e is the normalized value of the efficiency within the constraints; w_{tl}, w_{tm}, w_{tv}, and w_e are the weighting factors given to each index depending on the importance of the index to the users' requirements, and

$$w_{tl} + w_{tm} + w_{tv} + w_e = 1 \tag{2.32}$$

To normalize the indexes, unity-based normalization, denoted as feature scaling, is employed to bring all values into the range [0,1], as highlighted in eq. (2.33).

$$X' = \frac{X - X_{min}}{X_{max} - X_{min}} \tag{2.33}$$

where, X' is the normalized value, X is the original value, X_{min} is the minimum value of the dataset to be normalized, and X_{max} is the maximum value of the dataset to be normalized.

To determine the weighting factors in the objective function (2.31), analytic hierarchy process (AHP) is employed. The weighting factors w_{tl}, w_{tm}, w_{tv}, and w_e are calculated to 0.5001, 0.1915, 0.1915, and 0.1167, respectively .Detailed information about the calculation of the weighting factors is summarized in Appendix A. Hence, the final objective function is expressed as eq. (2.34).

$$f_{objective} = 0.5001 f_{tl} + 0.1915 f_{tm} + 0.1915 f_{tv} - 0.1167 f_e \tag{2.34}$$

2.4 Summary

From the analysis above, the design method of the superconducting DC generator can be summarized as follows. The design starts with the six control variables and generator specifications as the input parameters. The electromagnetic model is run to design the active parts of the generator which are taken the performance of the superconducting tapes and ferromagnetic materials into account. At the same time, the stresses on the rotor and stator components are calculated, such as radial stress, tangential stress and gravity. Consequently, the length of the superconducting tapes, the mass, the volume, and efficiency of the trial design are calculated. Optimisation is then used to evaluate the trial design according to objective function and decides the modifications of the control variables. A new trial design is continually made by repeating the same procedures until the optimal design is located. In the end, all design parameters of the generator result in the output. By this analytical design, a good computational efficiency is obtained. When the optimal design candidate is predicted by the design method, the next step is to use FEA software for further check, validate the design, and prepare for the development of the prototype in later chapter.

3 Electromagnetic model

To study the feasibility of superconducting DC machines, analytical analysis is usually preferred for a conceptual design because it is less time-consuming. In [131] an analytical method for the air-core and iron core generators was introduced and by multiplying factors the influence of iron in the rotor and stator was included. The magnetic field in the slotted structure with superconducting coils was calculated in [132] and in [133] a general analytical model for AC superconducting machines was proposed and magnetic flux density in different regions was calculated. Linear equations based on Laplace equations were used in [134] to calculate flux density by programing instead of analytical equations.Finally, in [9] field analysis to design a superconducting DC motor was used and its characteristics were analysed. To sum up, most of the previous work assumed the superconducting coils to be a current sheet. The critical current of a superconductor depends on the temperature, the magnetic field and the orientation of the magnetic field. Therefore, it is necessary to know the exact flux density distribution in the coils. The shape of the superconducting coils rather than a current sheet assumption ought to be taken into account in terms of flux density acting on the tapes. In [135] an analytical method with space harmonic analysis to calculate the flux density in the generator is put forward by employing magnetization vector. In [67] the flux density of the machines was analysed by the Laplace Equation. In [136] the magnetic field and design formulas were developed with air-cored superconducting machines. However, the influence of the iron core or the saturation in iron rotor/stator core is not considered so far. The magnetic equivalent circuit method is popular to design permanent magnet machines [137],[138].

In this section, a generalized and time-effective model for the superconducting DC generator is proposed in 2-dimension. The Poisson and Laplace Equation are employed for the magnetic field analysis.

3.1 Model description

In this model, as shown in Fig. 3.1, the shape of the superconducting coil is described by the inner radius of the superconducting coils r_{sci}, the outer radius of the superconducting coils r_{sco}, and the half angles between two adjacent coil sides α_1 and α_2. The inner and outer radiuses of the rotor and stator core are represented by r_{ri}, r_{ro}, r_{si}, and r_{so}, respectively. The angle θ is the mechanical angle. Region VI and region VII are air, which are outside the generator, region I is the rotor core, region II is the airgap winding, air gap, and inner dewar, region III is the superconducting coils, region IV is the outer dewar, and region V is the stator iron yoke. The cores of rotor and stator can be either made of ferromagnetic or non-magnetic materials. The nonlinearity of the ferromagnetic materials is also taken into consideration in this model. The end effects are neglected. The end effects can be included by empirical equations of the end winding in the design of the superconducting DC generator.

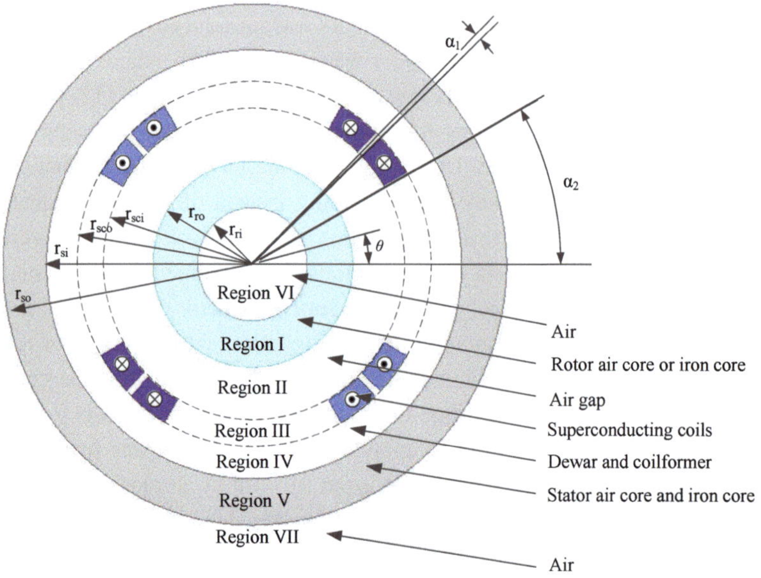

Fig. 3.1: General superconducting DC generator model.

3.2 Calculation of no-load flux density

3.2.1 Current density distribution

The most important part in the Poisson equation is the current source distribution. Therefore, the current density distribution of the superconducting coils is analysed firstly. The superconducting winding turn density distribution of one pole pair is shown in Fig. 3.2, where, T_{dsc0} is the winding turn density of the superconducting coil, which is assumed to be a constant value in this model and equals to the ratio of the turn number of a superconducting coil side to its cross section; p is the number of the pole pairs; θ is the mechanical angle; α_1 is the half angle between two adjacent coil sides of the different coils, α_2 is the half angle between two adjacent coil sides of the same coil; and the positive sign represents that the current is into the paper.

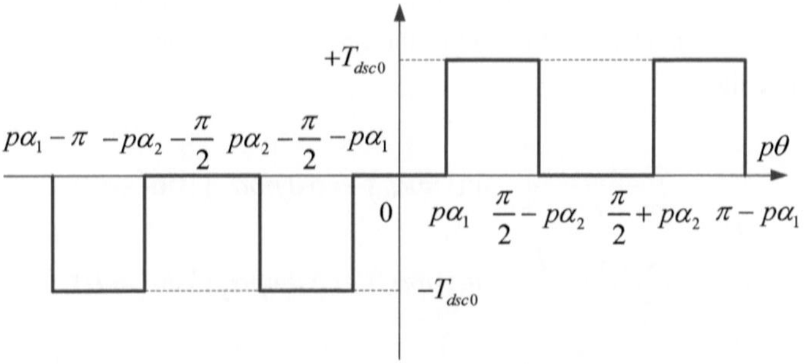

Fig. 3.2: Superconducting coils turns density distribution per pole pair.

The box function of Fig. 3.2 can be expressed by eq. (3.1) in (-π,π) region.

$$T_{dsc}(p\theta)=\begin{cases} +T_{dsc0} & \begin{array}{c} p\alpha_1 < p\theta < \frac{\pi}{2}-p\alpha_2, \\ \frac{\pi}{2}+p\alpha_2 < p\theta < \pi - p\alpha_1 \end{array} \\ -T_{dsc0} & \begin{array}{c} -\frac{\pi}{2}+p\alpha_2 < p\theta < -p\alpha_1, \\ -\pi + p\alpha_1 < p\theta < -\frac{\pi}{2}-p\alpha_2 \end{array} \\ 0 & else \end{cases} \tag{3.1}$$

where $T_{dsc}(p\theta)$ is the circumferential distribution of the winding turn density of the superconducting coils per pole pair.

With Fourier transform expansion, the winding turn density can be derived as

$$T_{dsc}(p\theta)=\begin{array}{l} \sum_{n=1}^{4t+1}\frac{4T_{dsc0}}{n\pi}\left[-\sin(np\alpha_1)+\cos(np\alpha_2)\right]\sin(np\theta) \\ +\sum_{n=3}^{4t+3}\frac{4T_{dsc0}}{n\pi}\left[\sin(np\alpha_1)+\cos(np\alpha_2)\right]\sin(np\theta) \end{array} \tag{3.2}$$

where n is the harmonic order, and t is an integer, which is 0, 1, 2, 3...

Then the circumferential current density distribution of the superconducting winding J_{scz} is calculated by eq. (3.3). The coefficients in the brackets are related to the shape of the superconducting coils.

$$J_{scz} = \begin{array}{l} \sum_{n=1}^{4t+1} \frac{4T_{dsc0} I_{sc}}{n\pi} \left[-\sin(np\alpha_1) + \cos(np\alpha_2) \right] \sin(np\theta) \\ + \sum_{n=3}^{4t+3} \frac{4T_{dsc0} I_{sc}}{n\pi} \left[\sin(np\alpha_1) + \cos(np\alpha_2) \right] \sin(np\theta) \end{array} \tag{3.3}$$

where I_{sc} is the current of the superconducting coils per turn, n is the harmonic order, and t is an integer, which is 0, 1, 2, 3...

3.2.2 Flux density generated by superconducting coils

In region I, II, IV,V,VI, and VII, magnetic vector potential φ has a z-direction component φ_z only, and the magnetic field can be described by the Laplace equation (3.4). In region III, the magnetic field can be described by the Poisson equation (3.5).

$$\frac{\partial^2 \varphi_{zm}}{\partial r^2} + \frac{1}{r}\frac{\partial \varphi_{zm}}{\partial r} + \frac{1}{r^2}\frac{\partial^2 \varphi_{zm}}{\partial \theta^2} = 0 \tag{3.4}$$

where m is the number of region, which equals to1, 2, 4, 5, 6, 7; and φ_{zm} is the magnetic vector potential of region m.

$$\frac{\partial^2 \varphi_{z3}}{\partial r^2} + \frac{1}{r}\frac{\partial \varphi_{z3}}{\partial r} + \frac{1}{r^2}\frac{\partial^2 \varphi_{z3}}{\partial \theta^2} = -\mu_0 J_{scz} \tag{3.5}$$

where φ_{z3} is the magnetic vector potential of region III, when m is 3; and μ_0 is the permeability of the vacuum, which is $4\pi\times10^{-7}$ H/m.

The z-direction component of magnetic vector potential φ_z in the superconducting DC generator is generated by the current density distribution of the superconducting coils J_{scz}. As a result, the general solutions of eq. (3.4) and (3.5) are as eq. (3.6) and (3.7), according to the symmetry of J_{scz}.

$$\varphi_{zm}(r,\theta) = \sum_{n=1}^{2t+1} \left(A_{nm} r^{np} + C_{nm} r^{-np} \right) \sin(np\theta) \tag{3.6}$$

$$\varphi_{z3}(r,\theta)=\begin{cases}\sum_{n}^{2t+1}\Big[\left(A_{n3}r^{np}+C_{n3}r^{-np}\right)\sin(np\theta) \\ \quad -\dfrac{k_e r^2}{4-(np)^2}\sin(np\theta)\Big] & np\neq 2 \\ \Big[\left(A_{n3}r^{2}+C_{n3}r^{-2}\right)\sin(2\theta) \\ \quad -\dfrac{1}{4}k_e r^2\ln r\sin(2\theta)\Big] & np=2\end{cases} \tag{3.7}$$

where, m is the region number, which is equal to 1, 2, 4, 5, 6, 7; r is the radius of the calculated point; n is the harmonic order; t is an integer, which is 1,2,3…; A_{nm} and C_{nm} are coefficients to express φ_z due to the nth harmonic in region m; A_{n3}, C_{n3} and k_e are coefficients to express φ_z due to the nth harmonic in region III when m is equal to 3. The coefficient k_e in eq. (3.7) is calculated by eq. (3.8).

$$k_e=\begin{matrix}\sum_{n=1}^{4t+1}\dfrac{4T_{dsc0}I_{sc}\mu_0}{n\pi}\left[-\sin(np\alpha_1)+\cos(np\alpha_2)\right] \\ +\sum_{n=3}^{4t+3}\dfrac{4T_{dsc0}I_{sc}\mu_0}{n\pi}\left[\sin(np\alpha_1)+\cos(np\alpha_2)\right]\end{matrix} \tag{3.8}$$

The tangential and radial component of flux density B_t, B_r can be calculated by the gradient of magnetic vector potential, as expressed in eq. (3.9) and (3.10), respectively.

$$\mathrm{B}_{tm}(r,\theta)=-\frac{\partial\varphi_{zm}}{\partial r} \tag{3.9}$$

$$\mathrm{B}_{rm}(r,\theta)=\frac{1}{r}\frac{\partial\varphi_{zm}}{\partial\theta} \tag{3.10}$$

where m=1,2,3,4,5,6,7.

By substituting eq. (3.6) and (3.7) into eq. (3.9) and (3.10), the flux density can be expressed as eq. (3.11) to (3.14).

$$\mathrm{B}_{tm}(r,\theta)=-\sum_{n=1}^{2t+1}\left(A_{nm}r^{np-1}-C_{nm}r^{-np-1}\right)np\sin(np\theta) \tag{3.11}$$

where, m is the region number, which is 1, 2, 4, 5, 6, 7; n is the harmonic order; and t is an integer, which is 1,2,3…

$$
\mathrm{B}_{t3}(r,\theta)=\begin{cases}
-\sum_{n=1}^{2t+1}\Big[\left(A_{n3}r^{np-1}-C_{n3}r^{-np-1}\right)np\sin(np\theta) & \\
\qquad -\dfrac{2k_e r}{4-(np)^2}\sin(np\theta)\Big] & np\neq 2\\
-\Big[2\left(A_{n3}r-C_{n3}r^{-3}\right)\sin(2\theta) & \\
-\dfrac{1}{2}k_e r\ln r\sin(2\theta)-\dfrac{1}{4}k_e r\ \sin(2\theta)\Big] & np=2
\end{cases} \tag{3.12}
$$

where, n is the harmonic order; and t is an integer, which is 1,2,3…

$$
\mathrm{B}_{rm}(r,\theta)=\sum_{n=1}^{2t+1}\left(A_{nm}r^{np-1}+C_{nm}r^{-np-1}\right)np\ \cos(np\theta) \tag{3.13}
$$

where, m is the region number, which is 1, 2, 4, 5, 6, 7; n is the harmonic order; and t is an integer, which is 1,2,3…

$$
B_{r3}(r,\theta)=\begin{cases}
\sum_{n=1}^{2t+1}\Big[\left(A_{n3}r^{np-1}+C_{n3}r^{-np-1}\right)np\cos(np\theta) & \\
\qquad -\dfrac{k_e r}{4-(np)^2}np\cos(np\theta)\Big] & np\neq 2\\
\Big[2\left(A_{n3}r+C_{n3}r^{-3}\right)\cos(2\theta) & \\
\qquad -\dfrac{1}{2}k_e r\ln r\cos(2\theta)\Big] & np=2
\end{cases} \tag{3.14}
$$

where, n is the harmonic order; and t is an integer, which is 1,2,3…

By employing proper boundary conditions between adjacent regions, the flux density in the generator can be obtained. The boundary conditions in the machine models are described as eq. (3.15).

The flux density in each region of the superconducting DC generator can be derived from eq. (3.11) to eq. (3.14) with the boundary conditions. As the materials in the stator core or rotor core can be ferromagnetic or nonmagnetic materials, different solutions are obtained. In order to clearly state the materials, the relative permeability is used. The relative permeability of the rotor core and stator is referred to as μ_{rr} and μ_{rs}, respectively. When relative permeability of the rotor core μ_{rr} is unequal to 1, the rotor core is made of iron; while relative permeability of the rotor core μ_{rr} is 1, the rotor core is made of nonmagnetic materials, so as to the stator. Coefficients A_{nm}, C_{nm}, A_{n3}, and C_{n3} are summarized as in Table 3.1, when both stator and rotor core are made of ferromagnetic materials. For other configuration, where only the stator core is made of ferromagnetic materials, or only the rotor core is made of ferromagnetic materials, or the stator and rotor core are made of nonmagnetic materials, are listed in Appendix B from Table A.B.1 to Table A.B.3, respectively.

$$\begin{cases} \varphi_{z6}(r,\theta)\big|_{r=0} = 0 \\ H_{t6}(r,\theta)\big|_{r=r_{ri}} = H_{t1}(r,\theta)\big|_{r=r_{ri}} \\ B_{r6}(r,\theta)\big|_{r=r_{ri}} = B_{r1}(r,\theta)\big|_{r=r_{ri}} \\ H_{t1}(r,\theta)\big|_{r=r_{ro}} = H_{t2}(r,\theta)\big|_{r=r_{ro}} \\ B_{r1}(r,\theta)\big|_{r=r_{ro}} = B_{r2}(r,\theta)\big|_{r=r_{ro}} \\ H_{t2}(r,\theta)\big|_{r=r_{sci}} = H_{t3}(r,\theta)\big|_{r=r_{sci}} \\ B_{r2}(r,\theta)\big|_{r=r_{sci}} = B_{r3}(r,\theta)\big|_{r=r_{sci}} \\ H_{t3}(r,\theta)\big|_{r=r_{sco}} = H_{t4}(r,\theta)\big|_{r=r_{sco}} \\ B_{r3}(r,\theta)\big|_{r=r_{sco}} = B_{r4}(r,\theta)\big|_{r=r_{sco}} \\ H_{t4}(r,\theta)\big|_{r=r_{si}} = H_{t5}(r,\theta)\big|_{r=r_{si}} \\ B_{r4}(r,\theta)\big|_{r=r_{si}} = B_{r5}(r,\theta)\big|_{r=r_{si}} \\ H_{t5}(r,\theta)\big|_{r=r_{so}} = H_{t7}(r,\theta)\big|_{r=r_{so}} \\ B_{r5}(r,\theta)\big|_{r=r_{so}} = B_{r7}(r,\theta)\big|_{r=r_{so}} \\ \varphi_{z7}(r,\theta)\big|_{r=\infty} = 0 \end{cases} \tag{3.15}$$

Table 3.1: Equations to calculate flux density when stator and rotor core are made of ferromagnetic materials

Coeff.	$np \neq 2$	$np=2$
K	$r_{ro}^{2np}\left(\frac{\mu_{rr}-1}{\mu_{rr}+1}\right)[1-(\frac{r_{ri}}{r_{ro}})^{2np}] \Big/ \left[1-\left(\frac{\mu_{rr}-1}{\mu_{rr}+1}\right)^2(\frac{r_{ri}}{r_{ro}})^{2np}\right]$	$r_{ro}^{4}\left(\frac{\mu_{rr}-1}{\mu_{rr}+1}\right)[1-(\frac{r_{ri}}{r_{ro}})^{4}] \Big/ \left[1-\left(\frac{\mu_{rr}-1}{\mu_{rr}+1}\right)^2(\frac{r_{ri}}{r_{ro}})^{4}\right]$
E	$r_{si}^{2np}[1-\left(\frac{\mu_{rs}-1}{\mu_{rs}+1}\right)^2(\frac{r_{si}}{r_{so}})^{2np}] \Big/ \left\{\left(\frac{\mu_{rs}-1}{\mu_{rs}+1}\right)[1-(\frac{r_{si}}{r_{so}})^{2np}]\right\}$	$r_{si}^{4}[1-\left(\frac{\mu_{rs}-1}{\mu_{rs}+1}\right)^2(\frac{r_{si}}{r_{so}})^{4}] \Big/ \left\{\left(\frac{\mu_{rs}-1}{\mu_{rs}+1}\right)[1-(\frac{r_{si}}{r_{so}})^{4}]\right\}$
A_{n2}	$\frac{1}{K-E}\frac{k_e(r_{sci}^{np+2}-r_{sco}^{np+2})}{2np(2+np)}+\frac{E}{K-E}\frac{k_e(r_{sci}^{2-np}-r_{sco}^{2-np})}{2np(2-np)}$	$\frac{1}{K-E}\frac{k_e(r_{sci}^{4}-r_{sco}^{4})}{16}+\frac{E}{K-E}\frac{k_e(\ln r_{sci}-\ln r_{sco})}{4}$
A_{n4}	$\frac{1}{K-E}\frac{k_e(r_{sci}^{np+2}-r_{sco}^{np+2})}{2np(2+np)}+\frac{K}{K-E}\frac{k_e(r_{sci}^{2-np}-r_{sco}^{2-np})}{2np(2-np)}$	$\frac{1}{K-E}\frac{k_e(r_{sci}^{4}-r_{sco}^{4})}{16}+\frac{K}{K-E}\frac{k_e(\ln r_{sci}-\ln r_{sco})}{4}$
C_{n2}	KA_{n2}	KA_{n2}
C_{n4}	EA_{n4}	EA_{n4}
A_{n3}	$A_{n2}+\frac{k_e r_{sci}^{2-np}}{2np(2-np)}$	$A_{n2}+\frac{k_e \ln r_{sci}}{4}+\frac{k_e}{16}$
C_{n3}	$C_{n2}-\frac{k_e r_{sci}^{np+2}}{2np(2+np)}$	$C_{n2}-\frac{k_e r_{sci}^{4}}{16}$
A_{n1}	$\frac{(\mu_{rr}+1)A_{n2}+(1-\mu_{rr})C_{n2}r_{ro}^{-2np}}{2}$	$\frac{(\mu_{rr}+1)A_{n2}+(1-\mu_{rr})C_{n2}r_{ro}^{-4}}{2}$
C_{n1}	$\frac{(1-\mu_{rr})A_{n2}r_{ro}^{2np}+(\mu_{rr}+1)C_{n2}}{2}$	$\frac{(1-\mu_{rr})A_{n2}r_{ro}^{4}+(\mu_{rr}+1)C_{n2}}{2}$
A_{n5}	$\frac{(\mu_{rs}+1)A_{n4}+(1-\mu_{rs})C_{n4}r_{si}^{-2np}}{2}$	$\frac{(\mu_{rs}+1)A_{n4}+(1-\mu_{rs})C_{n4}r_{si}^{-4}}{2}$
C_{n5}	$\frac{(1-\mu_{rs})A_{n4}r_{si}^{2np}+(\mu_{rs}+1)C_{n4}}{2}$	$\frac{(1-\mu_{rs})A_{n4}r_{si}^{4}+(\mu_{rs}+1)C_{n4}}{2}$
A_{n6}	$\frac{2A_{n1}}{\mu_{rr}+1}$	$\frac{2A_{n1}}{\mu_{rr}+1}$
C_{n7}	$\frac{2C_{n5}}{\mu_{rs}+1}$	$\frac{2C_{n5}}{(\mu_{rs}+1)}$
C_{n6}	0	0
A_{n7}	0	0

3.3 Iron effects and armature reaction

3.3.1 Iron core effects

To calculate the flux density more accurately, the nonlinearity of stator iron yoke or rotor iron yoke are taken into account and the *B-H* curve of ferromagnetic materials are included. If both the stator and rotor core are made of silicon laminations, the relative permeability of the rotor and stator core, u_{rr} and u_{rs}, determinations are not independent. Then an iterative algorithm is proposed to determine the relative permeability, as shown in Fig. 3.3, where u_{rr0}, u_{rs0} are the initial value of relative permeability in stator yoke and iron core, and u_{rr1}, u_{rs1} are the calculated value of relative permeability in stator yoke and iron core from *B-H* curve.

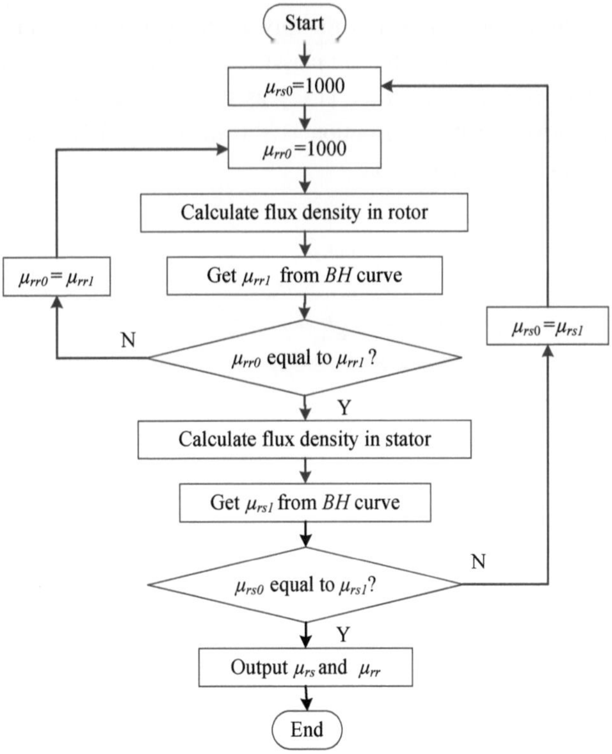

Fig. 3.3: Iterative algorithm to determine relative permeability.

As a start, an initial value of relative permeability in the rotor and stator core, u_{rr0} and u_{rs0}, are given. Then the flux density in the rotor core can be calculated and the relative permeability u_{rr1} through the *B-H* curve can be found. After that, the initial value u_{rr0} is compared to the calculated value u_{rr1}. If they are not equal, the initial value u_{rr0} is adjusted and the inner loop is calculated again until the initial value u_{rr0} and the calculated value u_{rr1} are equal. Then the flux density and the relative permeability in the stator core are calculated. Similarly, the initial value u_{rs0} and the calculated value u_{rs1} are compared and the outer loop is executed until the initial value u_{rs0} and the calculated value u_{rs1} are equal. At last, the appropriate relative permeability is found at each calculated mechanical angle θ for rotor and stator core, respectively. And for each mechanical angle θ, the calculated point in the rotor or stator core is located in its radially middle position.

3.3.2 Armature reaction

In order to consider the armature reaction, the flux density generated by the armature winding is also analysed. The slots in rotor are classified into slot pairs per pole pair and are numbered as slot pair 1/1', 2/2', ...*j*/*j*',... [132], as show in Fig. 3.4. For slot pair *j*/*j*', the armature winding turns density distribution is shown in Fig. 3.5. The box function of Fig. 3.5 can be expressed by eq. (3.16).

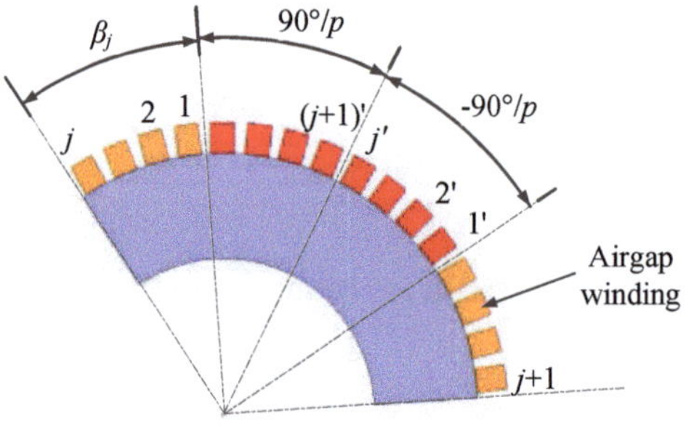

Fig. 3.4: Number of slot pair per pole pair.

$$T_{daj}(p\theta)=\begin{cases} +T_{da0} & \frac{\pi}{2}+p\beta_j-\frac{p\alpha_{sl}}{2}<p\theta<\frac{\pi}{2}+p\beta_j+\frac{p\alpha_{sl}}{2} \\ -T_{da0} & -\frac{\pi}{2}-p\beta_j-\frac{p\alpha_{sl}}{2}<p\theta<-\frac{\pi}{2}-p\beta_j+\frac{p\alpha_{sl}}{2} \end{cases} \tag{3.16}$$

where, T_{da0} is the winding turn density in each slot, which is a constant; p is the number of pole pair; α_{sl} is the slot spreading angle; and β_j is the mechanical angle shifting from slot pair 1, j=1,2,3...S, S is the number of slot pair per pole pair.

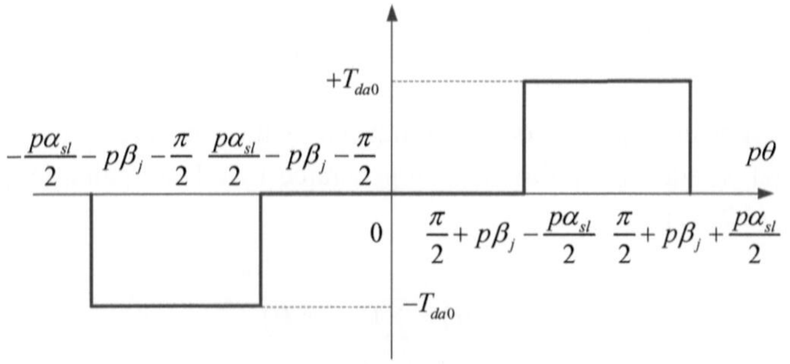

Fig. 3.5: Armature winding turns density of slot pair *j/j'* per pole pair.

With Fourier transform, the winding turn density of slot pair *j/j'* is as eq. (3.17).

$$T_{daj}\left(p\theta\right)=\begin{array}{l}-\sum_{n=1}^{4t+1}\frac{4T_{da0}}{n\pi}\sin\left(\frac{np\alpha_{sl}}{2}\right)\sin\left[np\left(\theta-\beta_j\right)\right]\\+\sum_{n=1}^{4t+3}\frac{4T_{da0}}{n\pi}\sin\left(\frac{np\alpha_{sl}}{2}\right)\sin\left[np\left(\theta-\beta_j\right)\right]\end{array}\tag{3.17}$$

where, $T_{daj}(p\theta)$ is the circumferential distribution of the winding turn density of slot pair *j/j'*, n is the nth harmonic order, and t is 1,2,3… .

The current density distribution of armature winding J_{aj} in *j/j'* slot pair is as eq. (3.18).

$$J_{aj}\left(p\theta\right)=\begin{array}{l}-\sum_{n=1}^{4t+1}\frac{4T_{da0}}{n\pi}i_a\sin\left(\frac{np\alpha_{sl}}{2}\right)\sin\left[np\left(\theta-\beta_j\right)\right]\\+\sum_{n=1}^{4t+3}\frac{4T_{da0}}{n\pi}i_a\sin\left(\frac{np\alpha_{sl}}{2}\right)\sin\left[np\left(\theta-\beta_j\right)\right]\end{array}\tag{3.18}$$

where i_a is the current per turn in armature winding.

The procedure to calculate the flux density generated by slot pair *j/j'* is the same as for superconducting coils. Equation (3.3) to (3.15) and Table 3.1, Table A.B.1 to Table A.B can be applied, except that the circumferential current density distribution of the superconducting winding J_{scz}, the current of the superconducting coils per turn I_{sc}, $\sin(np\theta)$, $\cos(np\theta)$, inner radius of the superconducting coils r_{sci}, outer radius of the superconducting coils r_{sco} and factor k_e are replaced by the current density distribution of the armature winding J_{aj}, the current per turn in armature winding i_a, $\sin[np(\theta-\beta_j)]$, $\cos[np(\theta-\beta_j)]$, the inner radius of the slots r_{sli}, the outer radius of the slots r_{slo} and factor k_{eaj} in eq. (3.19).

$$k_{eaj} = -\sum_{n=1}^{4t+1} \frac{4T_{da0}}{n\pi} i_a \sin\left(\frac{np\alpha_{sl}}{2}\right) + \sum_{n=1}^{4t+3} \frac{4T_{da0}}{n\pi} i_a \sin\left(\frac{np\alpha_{sl}}{2}\right) \tag{3.19}$$

Assuming superposition, the flux density produced by the armature winding per pole pair can be obtained by adding the flux density generated by all slot pairs per pole pair, expressed by eq. (3.20) and (3.21).

$$\mathrm{B}_{rm}(r,\theta) = \sum_{j=1}^{S} \mathrm{B}_{rmj}(r,\theta) \tag{3.20}$$

$$\mathrm{B}_{tm}(r,\theta) = \sum_{j=1}^{S} \mathrm{B}_{tmj}(r,\theta) \tag{3.21}$$

where, m represents the region number, which is 1, 2, 3, 4, 5, 6, 7; and j=1,2,3…S, S is the slot pair number per pole pair.

It needs to be pointed out that the region III represents the area of the armature winding instead of the superconducting coils in this section and the relative permeability of the rotor and stator core is from section 3.2.2. An air-gap winding is usually employed in the superconducting generators, however, for low speed direct-drive machines, the use of iron teeth armature winding can increase efficiency and reduce amount of superconducting wire [93]. To include the iron teeth in the general model, the Carter factor [33] is employed. On average, according to Carter's principle, the air gap seems to be longer than its physical airgap, compared to no iron teeth and slot on the rotor and stator surface. The effective gap can be expressed by eq. (3.22) [33].

$$\delta_{te} = \kappa_c \delta_{em} \tag{3.22}$$

where δ_{te} is the air gap that takes the iron teeth into account; δ_{em} is the effective electromagnetic air gap length, which takes the radial space for cryostat and coil former; κ_c is the Carter factor, which is expressed as eq. (3.23).

$$\kappa_c = \frac{\tau_u}{\tau_u - \kappa w_{sl}} \tag{3.23}$$

where τ_u is the slot pitch, w_{sl} is the slot width and κ can be calculated as eq. (3.24).

$$\kappa = \frac{2}{\pi}\left[\tan^{-1}\left(\frac{w_{sl}}{2\delta_{em}}\right) - \left(\frac{2\delta_{em}}{w_{sl}}\right)\ln\sqrt{1+\left(\frac{w_{sl}}{2\delta_{em}}\right)^2}\right] \tag{3.24}$$

3.4 Validation by finite element analysis

For the validation of the analytical method, the generator setup in Fig. 3.6 is used to compare the analytical model with results from finite element analysis (FEA) 2D *Maxwell*. The stator core and yoke core in the generator are made of ferromagnetic materials, which are the worst case in accuracy in the analytical method, since the nonlinearity of the iron would result in larger mismatch. The generator parameters used in this part are summarized in Table 3.2. Please note that the dimensions given in Table 3.2 are only a random example for a 10 kW demonstrator to validate the analytical method and are not for an optimized generator design. As stated in [3], superconducting generators are superior to permanent magnet generators for wind turbines with rated power of 8 MW or more. Hence, the air gap flux density in the 10 kW demonstrator are limited by the cost of the superconductor. The positions for comparing flux density in rotor core, air gap, superconducting coils, and stator core, which are B_1, B_2, B_3 and B_5, are shown in Fig. 3.6.

The magnetic flux density distribution under rated load can be calculated by summing the flux density under no load and the flux density generated by the armature winding, assuming superposition can be used. The results of the magnetic flux density in the air gap and superconducting coils under rated load are compared in Fig. 3.13 and Fig. 3.14. It can be seen that the analytical model seems feasible. The largest mismatch of radial component B_{r2} and B_{r3} is 7.2% and 17.4%. While the largest mismatch of tangential component B_{t2} and B_{t3} is 3.5% and 9.2%.

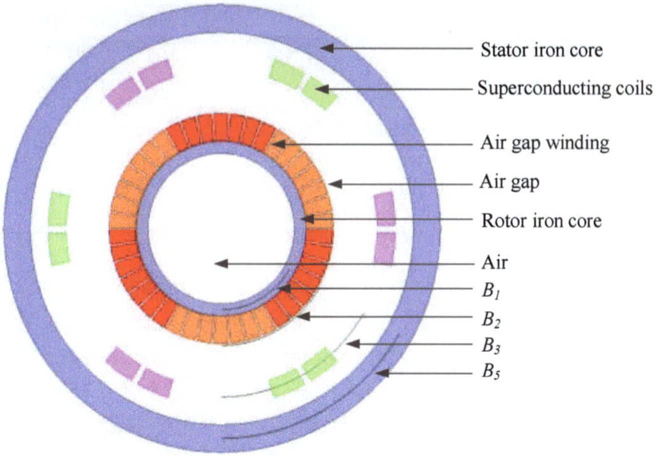

Fig. 3.6: Sketch of machine used in comparison between analytical and finite element analysis method.

By applying the equations in Table 3.1, tangential and radial components of flux density in rotor core B_1,air gap B_2, superconducting coils B_3, and stator core B_5 under no load are calculated, labelled as B_{t1}, B_{r1}, B_{t2}, B_{r2}, B_{t3}, B_{r3}, B_{t5}, B_{r5}. The analytical and FEA results are depicted in Fig. 3.7, Fig. 3.8, Fig. 3.9 and Fig. 3.10.The largest mismatch of radial component B_{r1}, B_{r2}, B_{r3}, and B_{r5} is 14.6%, 7.2%, 17.5%, and 19.2%, respectively. While the largest mismatch of tangential component B_{t1}, B_{t2}, B_{t3}, and B_{t5} is 3.6%, 3.4%, 9.1% and 8.2%. The mismatch of the two methods can be explained by the mismatches of operating permeability of iron cores. The analytical method calculates the relative permeability of points in the radially middle of the stator and rotor core, wheareas the FEA calculates the relative permeability of nodes in the whole stator and rotor core. The mismatches seem acceptable for an analytical calculation.

Table 3.2: Specifications of the superconducting DC generator

Symbol	Quantity	Value
P_N	Pole pair number	10 kW
p	Pole pair number	3
r_{ri}	Inner radius of the rotor core	81.6 mm
r_{ro}	Outer radius of the rotor core	94.6 mm
r_{sci}	Inner radius of the SC coil region	173.1 mm
r_{sco}	Outer radius of the SC coil region	193.1 mm
r_{si}	Inner radius of the stator core	213.1 mm
r_{so}	Outer radius of the stator core	243.1 mm
r_{sli}	Inner radius of the slot region	95.0 mm
r_{slo}	Outer radius of the slot region	124.8 mm
α_1	See Fig. 3.1	1.18 mechanical degree
α_2	See Fig. 3.1	18 mechanical degree
α_{sl}	Slot spreading angle	7.72 mechanical degree
I_{sc}	Current per SC turn	147.0 A
i_a	Current per armature winding turn	7.25 A
T_{dsc0}	Winding turn density per SC coil side	0.870 turn/mm^2
T_{da0}	Winding turn density per slot	0.172 turn/mm^2
Z_u	Slot number in total	42

The result of the flux density generated by the armature winding calculated by the analytical method and FEA are illustrated in Fig. 3.11 and Fig. 3.12. Only the flux densities in the physical air gap, B_2, and in the superconducting coils, B_3, are shown in the figures. The largest mismatch of radial component B_{r2} and B_{r3} is 2.9% and 8.7%. While the largest mismatch of tangential component B_{t2} and B_{t3} is 5.6% and 9.0%. The mismatches seem also acceptable for an analytical calculation. When the flux density generated by the armature winding is calculated with the analytical method, the relative permeability of the stator and rotor core under no load condition is employed. Moreover, in the analytical method, the relative permeability of the points in the radially middle of the stator and rotor core are calculated only instead of the whole stator and rotor core. Hence, the mismatches to calculate permeability of iron cores of the two different methods are the main reason in the flux density mismatches. It can be concluded that the the flux density generated by the air gap copper winding in the superconducting DC generator is relatively small. The reason is that the equivalent air gap in the generator is large.

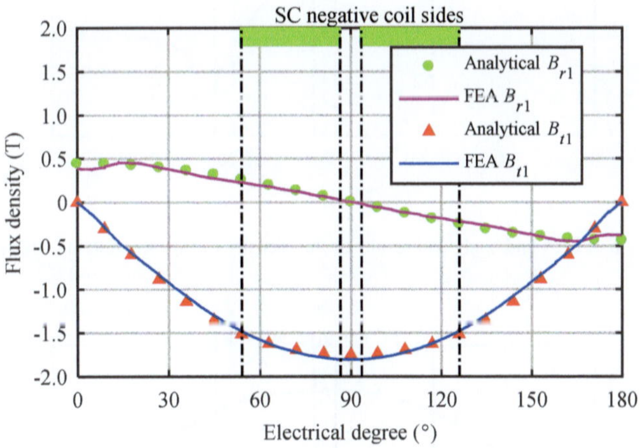

Fig. 3.7: Comparison of FEA and analytical results of flux density in rotor iron yoke under no load.

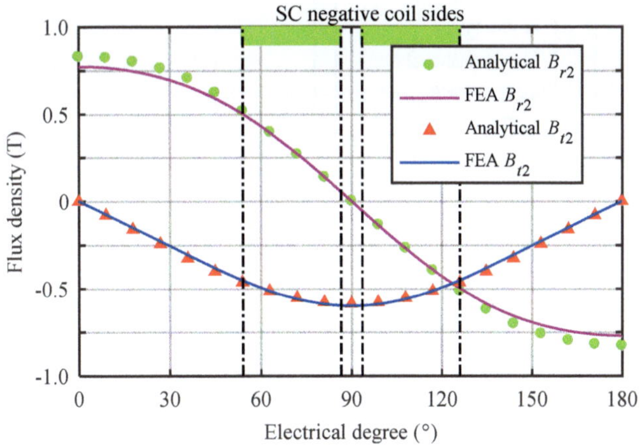

Fig. 3.8: Comparison of FEA and analytical results of flux density in physical air gap under no load.

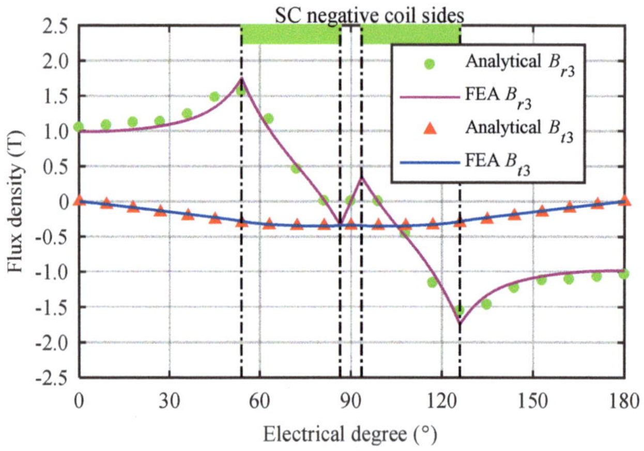

Fig. 3.9: Comparison of FEA and analytical results of flux density in superconducting coils under no load.

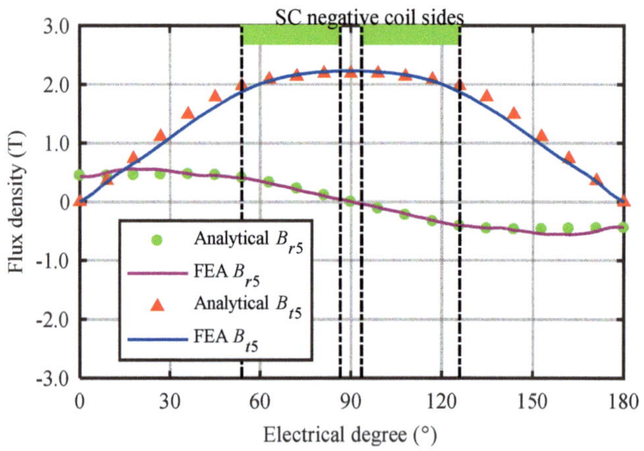

Fig. 3.10: Comparison of FEA and analytical results of flux density in stator iron yoke under no load.

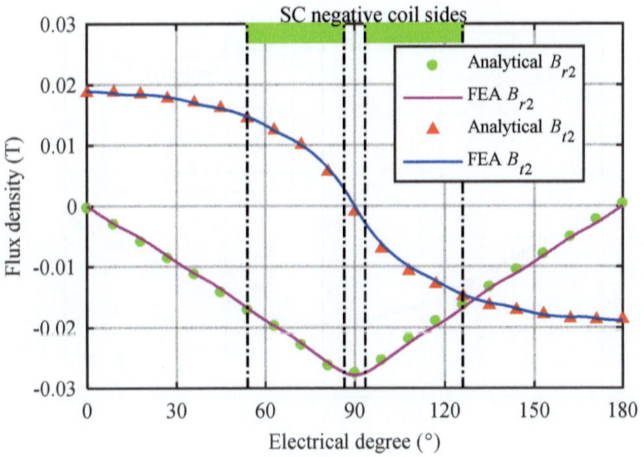

Fig. 3.11: Comparison of FEA and analytical results of the magnetic flux density generated by armature winding in the air gap.

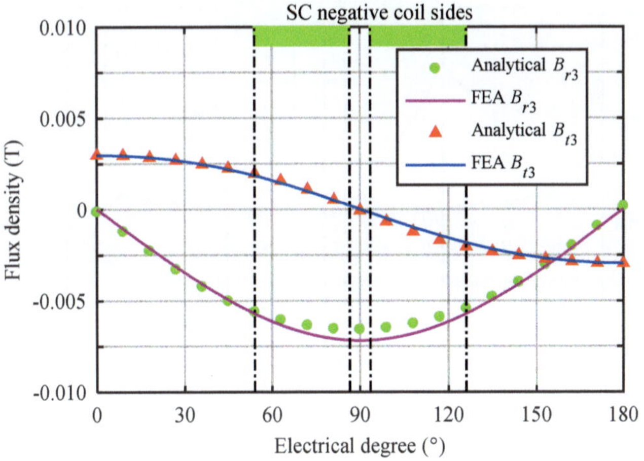

Fig. 3.12: Comparison of FEA and analytical results of the magnetic flux density generated by armature winding in the superconducting coils.

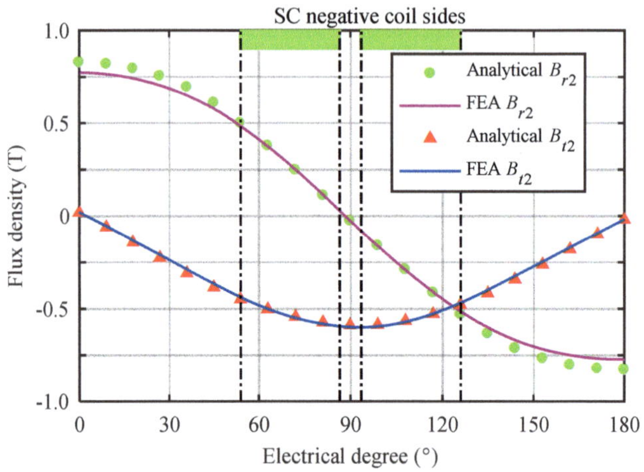

Fig. 3.13: Comparison of FEA and analytical results of the flux density under rated load in the air gap.

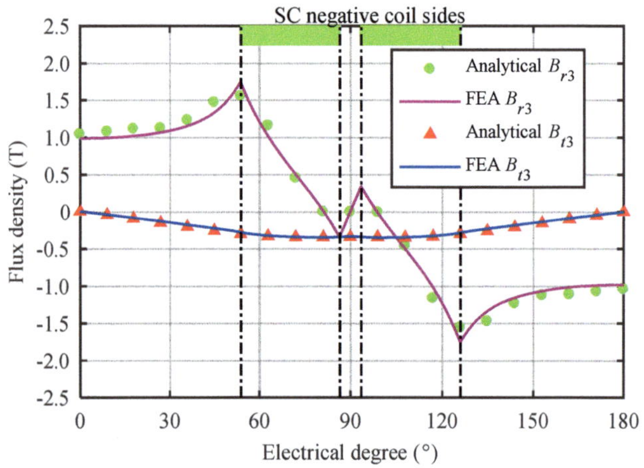

Fig. 3.14: Comparison of FEA and analytical results of the magnetic flux density under rated load in the superconducting coils.

3.5 Summary

The finite element analysis is used for the detailed calculation of the magnetic field inside the machine. The nonlinearity of the ferromagnetic materials is also included in the FEA, which results in accurate caculation. However, the calculation of a FEA model takes time, especially when fine meshed elements are employed to have a good flux density distribution. In the machine design process the number of variables are usually very high. The consumed time to find an optimised design will be very long in a multi-dimension searching space of variables. For instance, the computation of a 2D FEA model in section 3.7 with the static magnetic field calculation of *Maxwell* takes 47.3 seconds. If five variables are employed as explained in section 2.3, and each variable has only five different values within its range, then the computation time would be around 41 hours. In this case, the analytical method, including the nonlinearity of materials, takes only around 3.5 hours, with 4.0 seconds for each calculation. The comparison above excludes the searching process of the operating current of the superconducting tape. With an increase in the variables, the computation time will reach an unrealistic high value to design the machine within the total searching space with FEA, whereas the analytical method can save computation time. Moreover, the analytical model developed in Chapter 3 has a sufficient accuracy when compared to the FEA.

Due to the complexity of the variables and objectives of the generator design, an evolutionary algorithm is employed. Even though it is possible to implement the FEA into the optimisation process with the evolutionary algorithm [122],[139], high programming capability is required. The main objective in selecting a design method is to find a compromise between complexity, accuracy and speed. Hence, high efficiency can be obtained by combining the two methods. Firstly, the optimal design candidate is predicted by the analytical method and the evolutionary optimization algorithm. Then the design is further checked and modified by FEA method, including the commutation performance. After that, the final design is completed and detailed information of the generator is acquired.

After calculating the flux density distribution in the generator, the superconducting DC generator can be designed and its performance can be analysed. However, one limit in the analytical method is that the end winding is not included in the model. In the analytical model the cross section of the superconducting coils are not rectangular. The current carrying area (the cross section of the coil legs) can be described by radiuses and angles, r_{sci}, r_{sco}, α_1 and α_2. The 3D figure of such a superconducting coil is shown in Fig. 3.15(a). To manufacture such a coil the superconducting tapes need to bend. The superconducting coils are usually manufactured in pancakes, as shown in Fig. 3.15(b) and the cross section of the superconducting coils is in rectangular. The difference of the cross section in the analytical model and in practice is shown in Fig. 3.16.

The difference in the cross section of the superconducting coils will result in the fact that the calculated flux density in the analytical model is much larger. In a generator with a low pole pair number, the calculated flux density could be as large as 110% to 130%. To have a precise analytical model with rectangular cross section of superconducting coils would be too complicated. In this case the analytical model can only give a very rough generator design. The accuracy improves with higher pole pair number in the generator design. The reason is that the cross section of the superconducting coils approaches to the sector shape by increasing pole pair number as shown in Fig. 3.17.

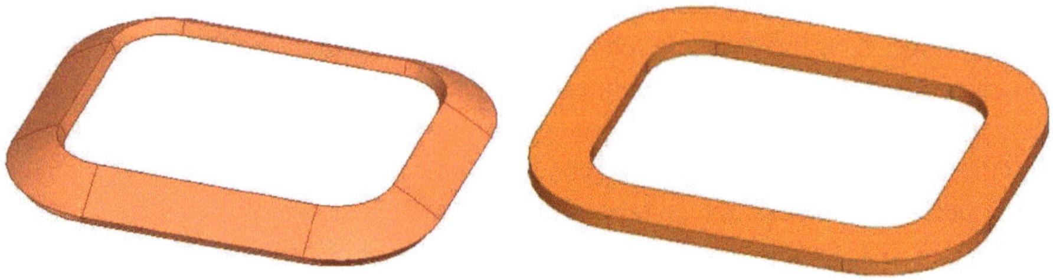

(a) (b)

Fig. 3.15: 3D superconducting coil in the analytical model and in practice, (a) analytical model, (b) in practice.

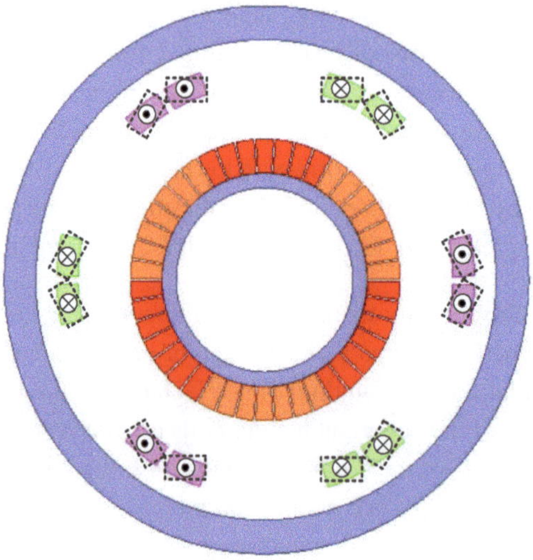

Fig. 3.16: Difference of the cross section in the analytical model and in practice, the dashed line is the cross section in practice.

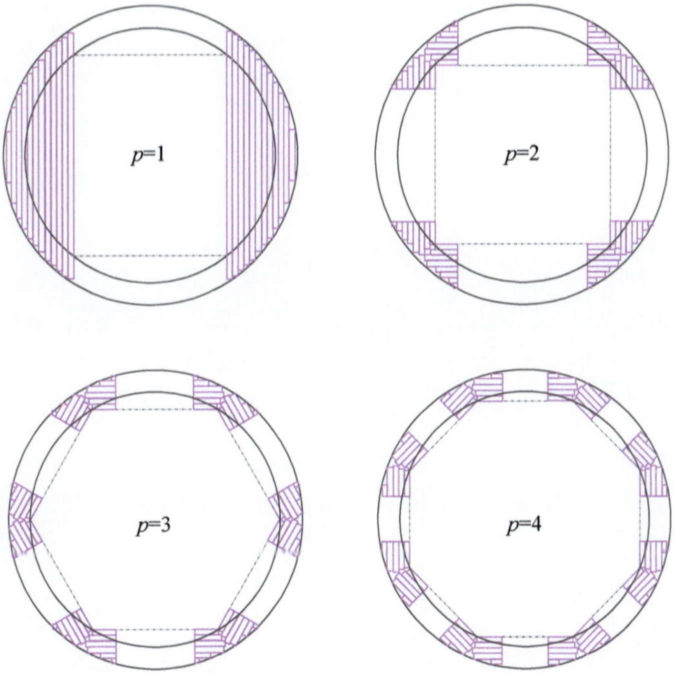

Fig. 3.17: Cross section of the superconducting coils with different coil layers and pole pairs.

4 Mechanical model

Applying the equations and process in Chapter 2 and 3, an optimized generator design can be obtained. In this design, only active materials are considered, which means that weight and volume are the sum of the stator iron core, superconducting wires, rotor iron core and copper wires. When the generator power becomes larger, neglecting inactive materials may not be optimal anymore. From the results of the permanent magnet generator design [56] it can be seen that the influence of the structure materials on weight increases sharply when the generator output power increases, and they even can be the dominant weight. Hence, it is necessary to take into consideration the mechanical structure in the optimisation process.

4.1 Force analysis in the generator

The shear stress is useful in an electrical machine to calculate the electromagnetic torque. In Fig. 4.1 the direction of the shear stress in the superconducting wind generator is illustrated. If saturation of the stator and rotor iron core is not considered, the general equation to calculate the shear stress in an electrical machine is expressed in eq. (4.1).

$$T = \frac{\pi \bar{\sigma} D_a^2 l}{2} \tag{4.1}$$

where T is the rated torque, $\bar{\sigma}$ is the shear stress, D_a is the armature diameter, and l is the stack length of the electric machine.

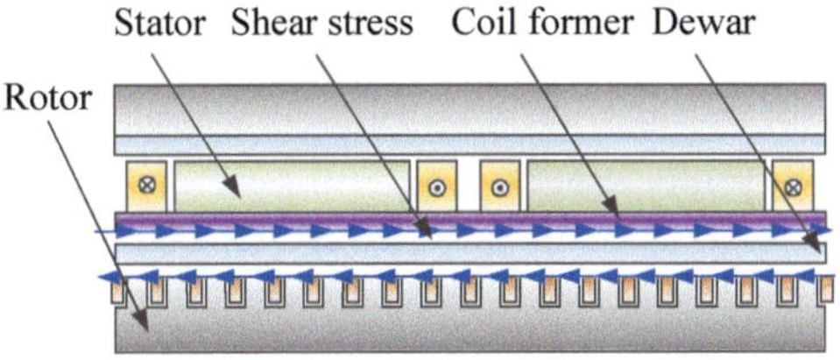

Fig. 4.1: Shear stress of the superconducting generator.

A normal force is directed across the airgap and tries to attract the stator and rotor together. The direction of the normal stress is shown in Fig. 4.2 in a superconducting generator. The general equation to calculate the normal stress is expressed in eq. (4.2).

$$\overline{q} = \hat{B}^2 / (2\mu_0) \tag{4.2}$$

where $\overline{q}$ is the normal stress, $\hat{B}$ is the fundamental radial magnitude, and u_0 is the permeability of free space.

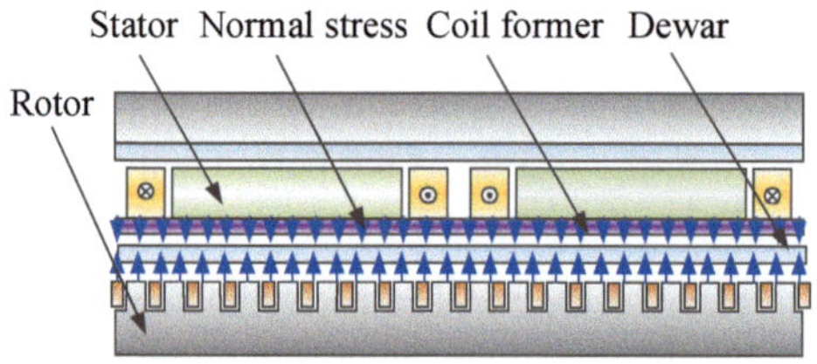

Fig. 4.2: Normal stress of the superconducting wind generator.

In wind turbines, the axis of the rotor blades and the generator is tilted by an angle, φ, of 5-10 degree to avoid the tower from being struck by the blades. Hence, there will be deflection of the stator and rotor due to gravity, as illustrated in Fig. 4.3.

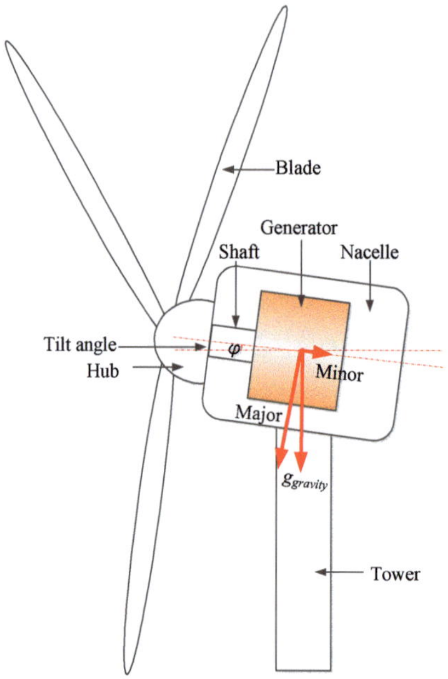

Fig. 4.3: Sketch of the tilt angle and gravity components of the superconducting generator.

4.2 Deflection calculation

The analytical model for the mechanical design of a permanent magnet generator has been developed by [140]-[142]. As the shear stress to generate the torque has no influence in axial and radial deflection, the shear stress is not considered. For the major gravity componeny, the total deflection of the stator and rotor will be nearly zero as the deflection of the stator and rotor are in the same direction, if same stiffness and materials are used in the stator and rotor. Hence, only the influence of the minor component is considered. For simplicity, the rotor and stator support structure is illustrated in Fig. 4.4, and no weight reduction method is adopted.

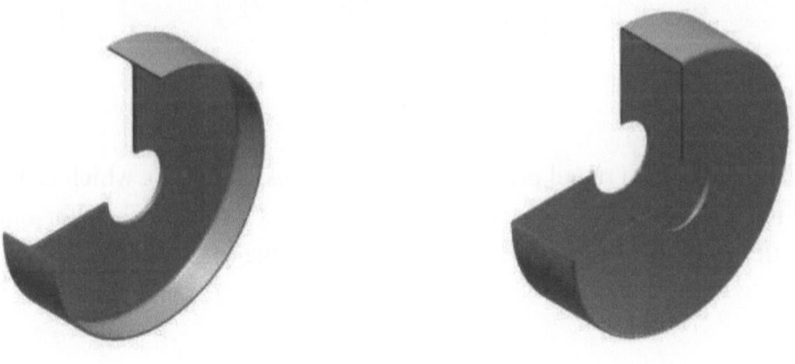

(a) (b)

Fig. 4.4: Support structure of the generator, (a) rotor support, and (b) stator support.

4.2.1 Radial deflection of rotor back iron

The radial deflection, u_{rdflr}, of the rotor by disc support [141] is calculated as follows.

$$u_{rdflr} = \frac{q_{Nstress} r_{ri}^2}{E_{steel} h_{ry}} + \frac{f_{rdisc}}{2D_{coeffry}\lambda_{coeffry}^3} \left[\begin{array}{l} \frac{-F_1(x=0)}{C_{11}}(C_3C_{a2} - C_4C_{a1}) \\ + \frac{F_2(x=0)}{2C_{11}}(C_2C_{a2} - 2C_3C_{a1}) - \frac{F_{a4}(x=0)}{2} \end{array} \right] \tag{4.3}$$

where, u_{rdflr} is the maximum radial deflection at both ends of the rotor, $q_{Nstress}$ is the uniform normal stress, h_{ry} is the rotor's back yoke thickness, r_{ri} is the inner radius of the rotor's back yoke, E_{steel} is the Young's modulus for steel, f_{rdisc} is the uniform outward radial force per unit length around the outside of the disc due to the normal stress $q_{Nstress}$, $D_{coeffry}$, $\lambda_{coeffry}$, C_{11},C_2, C_3,C_4,C_{a2},C_{a1}, F_1, F_2, F_{a4} are coefficients used in calculation, which are calculated as follows, as well as f_{rdisc}.

$$f_{rd} = \frac{q_{Nstress} r_{ri}^2}{E_{steel} h_{ry} \left[\frac{r_{ri}}{E_{steel}\left(r_{ri}^2 - r_{shaft}^2\right)} \left[\left(1-\nu_{steel}\right) r_{ri}^2 + \left(1+\nu_{steel}\right) r_{shaft}^2 \right] - t_{rdisc} \left[\begin{array}{l} \frac{1}{4D_{coeffry}\lambda_{coeffry}^3} \frac{C_2 C_{a2} - 2C_3 C_{a1}}{C_{11}} F_2\left(x = {}^{l_{ef}}/_{2}\right) \\ -\frac{1}{2D_{coeffry}\lambda_{coeffry}^3} \frac{C_3 C_{a2} - C_4 C_{a1}}{C_{11}} F_1\left(x = {}^{l_{ef}}/_{2}\right) \\ -\frac{1}{4D_{coeffry}\lambda_{coeffry}^3} F_{a4}\left(x = {}^{l_{ef}}/_{2}\right) \end{array} \right] \right]} \tag{4.4}$$

$$f_{rdisc} = t_{rdisc} f_{rd} \tag{4.5}$$

where, f_{rd} is the outward radial pressure, ν_{steel} is the Poisson's ratio, which is 0.3, r_{shaft} is the radius of the rotor's shaft, which we assume the same as the stator shaft for simplification; l_{ef} is the effective stack length, t_{rdisc} is the thickness of the rotor disc.

$$\lambda_{coeffry} = \left[\frac{3\left(1-\nu_{steel}^2\right)}{R_{ry}^2 h_{ry}^2} \right]^{1/4} \tag{4.6}$$

$$R_{ry} = \frac{r_{ri} + r_{ro}}{2} \tag{4.7}$$

where R_{ry} is the radius of the rotor structure, r_{ri} is the inner radius of the rotor's back yoke, and r_{ro} is the outer radius of the rotor's back yoke.

$$D_{coeffry} = \frac{E_{steel} h_{ry}^3}{12\left(1-\nu_{steel}^2\right)} \tag{4.8}$$

$$C_2 = \cosh\left(\lambda_{coeffry} l_{ef}\right) \sin\left(\lambda_{coeffry} l_{ef}\right) + \sinh\left(\lambda_{coeffry} l_{ef}\right) \cos\left(\lambda_{coeffry} l_{ef}\right) \tag{4.9}$$

$$C_3 = \sinh\left(\lambda_{coeffry} l_{ef}\right) \sin\left(\lambda_{coeffry} l_{ef}\right) \tag{4.10}$$

$$C_4 = \cosh\left(\lambda_{coeffry} l_{ef}\right) \sin\left(\lambda_{coeffry} l_{ef}\right) - \sinh\left(\lambda_{coeffry} l_{ef}\right) \cos\left(\lambda_{coeffry} l_{ef}\right) \tag{4.11}$$

$$C_{11} = \sinh^2\left(\lambda_{coeffry} l_{ef}\right) - \sin^2\left(\lambda_{coeffry} l_{ef}\right) \tag{4.12}$$

$$\begin{aligned} C_{a1} &= \cosh\left(\lambda_{coeffry}\left(l_{ef} - a\right)\right)\cos\left(\lambda_{coeffry}\left(l_{ef} - a\right)\right) \\ &= \cosh\left(\lambda_{coeffry} l_{ef}/2\right)\cos\left(\lambda_{coeffry} l_{ef}/2\right) \end{aligned} \tag{4.13}$$

$$\begin{aligned} C_{a2} &= \cosh\left(\lambda_{coeffry}\left(l_{ef} - a\right)\right)\sin\left(\lambda_{coeffry}\left(l_{ef} - a\right)\right) \\ &\quad + \sinh\left(\lambda_{coeffry}\left(l_{ef} - a\right)\right)\cos\left(\lambda_{coeffry}\left(l_{ef} - a\right)\right) \\ &= \cosh\left(\lambda_{coeffry} l_{ef}/2\right)\sin\left(\lambda_{coeffry} l_{ef}/2\right) \\ &\quad + \sinh\left(\lambda_{coeffry} l_{ef}/2\right)\cos\left(\lambda_{coeffry} l_{ef}/2\right) \end{aligned} \tag{4.14}$$

$$F_1(x) = \cosh\left(\lambda_{coeffry} x\right)\cos\left(\lambda_{coeffry} x\right) \tag{4.15}$$

$$F_2(x) = \cosh\left(\lambda_{coeffry} x\right)\sin\left(\lambda_{coeffry} x\right) + \sinh\left(\lambda_{coeffry} x\right)\cos\left(\lambda_{coeffry} x\right) \tag{4.16}$$

$$\begin{aligned} F_{a4}(x) &= \cosh\left(\lambda_{coeffry}\langle x - a\rangle\right)\sin\left(\lambda_{coeffry}\langle x - a\rangle\right) \\ &\quad - \sinh\left(\lambda_{coeffry}\langle x - a\rangle\right)\cos\left(\lambda_{coeffry}\langle x - a\rangle\right) \end{aligned} \tag{4.17}$$

where the use of the angle brackets $\langle\ \rangle$ is defined as follows: If $x < a$, $\langle x-a\rangle = 0$; if $x > a$, $\langle x-a\rangle = 1$. At $x=a$, the unit step function $\langle x-a\rangle$ is undefined just as vertical shear is undefined directly beneath a concentrated load.

4.2.2 Radial deflection of stator back iron

The radial deflection, u_{rdfls}, of the stator by disc support [141] is calculated as follows.

$$u_{rdfls} = \frac{q_{Nstress} r_{si}^2}{E_{steel} h_{sy}} - \frac{f_{sdisc}}{D_{coeffsy} \lambda_{coeffsy}^3} \left[\frac{C_{14}}{2C_{11}} F_2\left(x = l_{ef}/2\right) - \frac{C_{13}}{C_{11}} F_1\left(x = l_{ef}/2\right) - \frac{1}{2} F_4\left(x = l_{ef}/2\right) \right] \tag{4.18}$$

where u_{rdfls} is the radial deflection of the stator structure due to the normal stress, $q_{Nstress}$ is the uniform normal stress, h_{sy} is the stator's back yoke thickness, r_{si} is the inner radius of the stator's back yoke, E_{steel} is the Young's modulus for steel, f_{sdisc} is the uniform inward radial force per unit length around the outside of the disc due to the normal stress $q_{Nstress}$, $D_{coeffsy}$, $\lambda_{coeffsy}$, C_{11},C_{13}, C_{14}, F_1, F_2, F_4 are coefficients used in calculation, which are calculated as follows, as well as f_{sdisc}.

$$f_{sd} = \frac{q_{Nstress} r_{si}^2}{E_{steel} h_{sy} \left[\frac{r_{si}}{E_{steel}\left(r_{si}^2 - r_{shaft}^2\right)} \left[(1 - v_{steel}) r_{si}^2 + (1 + v_{steel}) r_{shaft}^2 \right] + \frac{t_{sdisc}}{2 D_{coeffsy} \lambda_{coeffsy}^3} \left[\frac{C_{14}}{2C_{11}} \begin{pmatrix} F_2(x=0) \\ +F_2(x=l_{ef}) \end{pmatrix} - \frac{C_{13}}{C_{11}} \begin{pmatrix} F_1(x=0) \\ +F_1(x=l_{ef}) \end{pmatrix} - \frac{1}{2} \begin{pmatrix} F_4(x=0) \\ +F_4(x=l_{ef}) \end{pmatrix} \right] \right]} \tag{4.19}$$

$$f_{sdisc} = t_{sdisc} f_{sd} \tag{4.20}$$

where, f_{sd} is the inward radial pressure, v_{steel} is the poisson's ratio, which is 0.3, r_{shaft} is the radius of the stator's shaft, which we assume the same as the rotor shaft for simplification; l_{ef} is the effective stack length, t_{sdisc} is the thickness of the stator disc.

$$\lambda_{coeffsy} = \left[\frac{3\left(1 - v_{steel}^2\right)}{R_{sy}^2 h_{sy}^2} \right]^{1/4} \tag{4.21}$$

$$R_{sy} = \frac{r_{si} + r_{so}}{2} \tag{4.22}$$

where R_{sy} is the radius of the stator structure, r_{si} is the inner radius of the stator's back yoke, and r_{so} is the outer radius of the stator's back yoke.

$$D_{coeffsy} = \frac{E_{steel} h_{sy}^3}{12\left(1 - v_{steel}^2\right)} \tag{4.23}$$

$$C_{11} = \sinh^2\left(\lambda_{coeffsy} l_{ef}\right) - \sin^2\left(\lambda_{coeffsy} l_{ef}\right) \tag{4.24}$$

$$C_{13} = \cosh\left(\lambda_{coeffsy} l_{ef}\right) \sinh\left(\lambda_{coeffsy} l_{ef}\right) - \cos\left(\lambda_{coeffsy} l_{ef}\right) \sin\left(\lambda_{coeffsy} l_{ef}\right) \tag{4.25}$$

$$C_{14} = \sinh^2\left(\lambda_{coeffsy} l_{ef}\right) + \sin^2\left(\lambda_{coeffsy} l_{ef}\right) \tag{4.26}$$

$$F_1(x) = \cosh\left(\lambda_{coeffsy} x\right)\cos\left(\lambda_{coeffsy} x\right) \tag{4.27}$$

$$F_2(x) = \cosh\left(\lambda_{coeffsy} x\right)\sin\left(\lambda_{coeffsy} x\right) + \sinh\left(\lambda_{coeffsy} x\right)\cos\left(\lambda_{coeffsy} x\right) \tag{4.28}$$

$$F_4(x) = \cosh\left(\lambda_{coeffsy} x\right)\sin\left(\lambda_{coeffsy} x\right) - \sinh\left(\lambda_{coeffsy} x\right)\cos\left(\lambda_{coeffsy} x\right) \tag{4.29}$$

4.2.3 Axial deflection of rotor back iron

The axial deflection of the rotor disc, u_{adflr}, is [141],[142]

$$u_{adflr} = \frac{-W_{ry} r_{ri}^4}{r_{shaft} D_{coeffrd}}\left(\frac{C_2 C_6}{C_5} - C_3\right) + M_{rb}\frac{r_{ri}^2}{D_{coeffrd}} C_2 + Q_b \frac{r_{ri}^3}{D_{coeffrd}} C_3 - \frac{w_{rdisc} r_{ri}^4}{D_{coeffrd}} L_{11} \tag{4.30}$$

where u_{adflr} is the axial deflection of the rotor due to the gravity, W_{ry} is the line load due to the weight of the rotor back yoke, w_{rdisc} is the uniformly distributed pressure due to the weight of the rotor disc, r_{ri} is the inner radius of rotor's back yoke, r_{shaft} is the radius of rotor's shaft, coefficients M_{rb}, Q_b, C_2, C_3, C_5, C_6, L_{11}, L_{14}, $D_{coeffrd}$ are calculated as follows.

$$W_{ry} = \rho_{steel} g_{gravity} \sin(\varphi) l_{ef} h_{ry} \tag{4.31}$$

$$w_{rdisc} = \rho_{steel} g_{gravity} \sin(\varphi) t_{rdisc} \tag{4.32}$$

where, $g_{gravity}$ is the gravity acceleration, ρ_{steel} is the mass density of the steel, and φ is the tilt angle of the generator to the horizontal axis,

$$M_{rb} = \frac{-w_{rdisc} r_{ri}^2}{C_5}\left[\frac{C_6}{2 r_{ri} r_{shaft}}\left(r_{ri}^2 - r_{shaft}^2\right) - L_{14}\right] \tag{4.33}$$

$$Q_b = \frac{w_{rdisc}}{2 r_{shaft}}\left(r_{ri}^2 - r_{shaft}^2\right) \tag{4.34}$$

$$L_{14}=\frac{1}{16}\left[1-\left(\frac{r_{shaft}}{r_{ri}}\right)^4-4\left(\frac{r_{shaft}}{r_{ri}}\right)^2\ln\frac{r_{ri}}{r_{shaft}}\right] \tag{4.35}$$

$$C_2=\frac{1}{4}\left[1-\left(\frac{r_{shaft}}{r_{ri}}\right)^2\left(1+2\ln\left(\frac{r_{ri}}{r_{shaft}}\right)\right)\right] \tag{4.36}$$

$$C_3=\frac{r_{shaft}}{4r_{ri}}\left\{\left[\left(\frac{r_{shaft}}{r_{ri}}\right)^2+1\right]\ln\left(\frac{r_{ri}}{r_{shaft}}\right)+\left(\frac{r_{shaft}}{r_{ri}}\right)^2-1\right\} \tag{4.37}$$

$$C_5=\frac{1}{2}\left[1-\left(\frac{r_{shaft}}{r_{ri}}\right)^2\right] \tag{4.38}$$

$$C_6=\frac{r_{shaft}}{4r_{ri}}\left[\left(\frac{r_{shaft}}{r_{ri}}\right)^2-1+2\ln\left(\frac{r_{ri}}{r_{shaft}}\right)\right] \tag{4.39}$$

$$L_{11}=\frac{1}{64}\left\{1+4\left(\frac{r_{shaft}}{r_{ri}}\right)^2-5\left(\frac{r_{shaft}}{r_{ri}}\right)^4-4\left(\frac{r_{shaft}}{r_{ri}}\right)^2\left[2+\left(\frac{r_{shaft}}{r_{ri}}\right)^2\right]\ln\left(\frac{r_{ri}}{r_{shaft}}\right)\right\} \tag{4.40}$$

$$D_{coeffrd}=\frac{E_{steel}t_{rdisc}^3}{12\left(1-v_{steel}^2\right)} \tag{4.41}$$

4.2.4 Axial deflection of stator back iron

The axial deflection of the stator disc,u_{adfls},is [141],[142]

$$u_{adfls}=\frac{-W_{sy}r_{si}^4}{r_{shaft}D_{coeffsd}}\left(\frac{C_2C_6}{C_5}-C_3\right)+M_{rb}\frac{r_{si}^2}{D_{coeffsd}}C_2+Q_b\frac{r_{si}^3}{D_{coeffsd}}C_3-\frac{w_{sdisc}r_{si}^4}{D_{coeffsd}}L_{11} \tag{4.42}$$

where u_{adfls} is the axial deflection of the stator disc due to the gravity, W_{sy} is the line load due to the weight of the stator back yoke, w_{sdisc} is the uniformly distributed pressure due to the weight of the stator disc, r_{si} is the inner radius of the stator's back yoke, r_{shaft} is the radius of the stator's shaft, the coefficients M_{rb}, Q_b, C_2, C_3, C_5, C_6, L_{11}, $D_{coeffsd}$ are calculated through the following equations.

$$W_{sy} = \rho_{steel} g_{gravity} \sin(\varphi)\left(l_{ef} - 2t_{sdisc}\right) h_{sy} \tag{4.43}$$

$$w_{sdisc} = \rho_{steel} g_{gravity} \sin(\varphi) t_{sdisc} \tag{4.44}$$

where, $g_{gravity}$ is the gravity acceleration, ρ_{steel} is the mass density of the steel, and φ is the tilt angle of the generator to the horizontal axis, t_{sdisc} is the thickness of the stator disc.

$$M_{rb} = \frac{-w_{sdisc} r_{si}^2}{C_5}\left[\frac{C_6}{2 r_{si} r_{shaft}}\left(r_{si}^2 - r_{shaft}^2\right) - L_{14}\right] \tag{4.45}$$

$$Q_b = \frac{w_{sdisc}}{2 r_{shaft}}\left(r_{si}^2 - r_{shaft}^2\right) \tag{4.46}$$

$$L_{14} = \frac{1}{16}\left[1 - \left(\frac{r_{shaft}}{r_{si}}\right)^4 - 4\left(\frac{r_{shaft}}{r_{si}}\right)^2 \ln\frac{r_{si}}{r_{shaft}}\right] \tag{4.47}$$

$$C_2 = \frac{1}{4}\left[1 - \left(\frac{r_{shaft}}{r_{si}}\right)^2\left(1 + 2\ln\left(\frac{r_{si}}{r_{shaft}}\right)\right)\right] \tag{4.48}$$

$$C_3 = \frac{r_{shaft}}{4 r_{si}}\left\{\left[\left(\frac{r_{shaft}}{r_{si}}\right)^2 + 1\right]\ln\left(\frac{r_{si}}{r_{shaft}}\right) + \left(\frac{r_{shaft}}{r_{si}}\right)^2 - 1\right\} \tag{4.49}$$

$$C_5 = \frac{1}{2}\left[1 - \left(\frac{r_{shaft}}{r_{si}}\right)^2\right] \tag{4.50}$$

$$C_6 = \frac{r_{shaft}}{4 r_{si}}\left[\left(\frac{r_{shaft}}{r_{si}}\right)^2 - 1 + 2\ln\left(\frac{r_{si}}{r_{shaft}}\right)\right] \tag{4.51}$$

$$L_{11} = \frac{1}{64}\left\{1 + 4\left(\frac{r_{shaft}}{r_{si}}\right)^2 - 5\left(\frac{r_{shaft}}{r_{si}}\right)^4 - 4\left(\frac{r_{shaft}}{r_{si}}\right)^2\left[2 + \left(\frac{r_{shaft}}{r_{si}}\right)^2\right]\ln\left(\frac{r_{si}}{r_{shaft}}\right)\right\} \tag{4.52}$$

$$D_{coeffsd} = \frac{E_{steel} t_{sdisc}^3}{12\left(1 - v_{steel}^2\right)} \tag{4.53}$$

4.3 Validation by finite element analysis

To verify the analytical method, the maximum deflection of the rotor and stator structural support due to the normal stress and the gravity is calculated, respectively, both by the analytical model and FE method. The summary of the parameters used is illustrated in Table 4.1.The physical air gap in this example is 10 mm. According to the analytical model, the maximum deflection of the stator support due to normal stress and gravity is 0.63 mm and 0.49 mm. The maximum deflection of the rotor support due to normal stress and gravity is 0.54 mm and 0.61 mm. The maximum deflection calculated by FE model compared to analytical results is shown in Fig. 4.5. From the results it can be seen that the mismatches seem still acceptable between FEA and analytical results.

Table 4.1: Summary of parameters used in FEA validation

Symbol	Explanation	Unit
$q_{Nstress}$	Uniform normal stress	3.2 MN/m^2
E_{steel}	Young's modulus for steel	210 GPa
r_{ri}	Inner radius of the rotor's back yoke	2455 mm
r_{ro}	Outer radius of the rotor's back yoke	2650 mm
h_{ry}	Rotor's back yoke thickness	195 mm
R_{ry}	Radius of the rotor structure	2553 mm
r_{shaft}	Radius of the shaft	1250 mm
l_{ef}	Effective stack length	1332 mm
l_{end}	Axial length of end armature winding at one end	305 mm
t_{rdisc}	Thickness of the rotor disc support	35 mm
r_{si}	Inner radius of the stator's back yoke	2823 mm
r_{so}	Outer radius of the stator's back yoke	3017 mm
h_{sy}	Stator's back yoke thickness	194 mm
R_{sy}	Radius of the stator structure	2920 mm
t_{sdisc}	Thickness of the stator disc support	45 mm
$g_{gravity}$	Gravity acceleration	9.81 N/kg
ρ_{steel}	Mass density of the steel	7850 kg/m^3
φ	Tilt angle of the generator to the horizontal axis	5°
δ	Physical air-gap length	10 mm

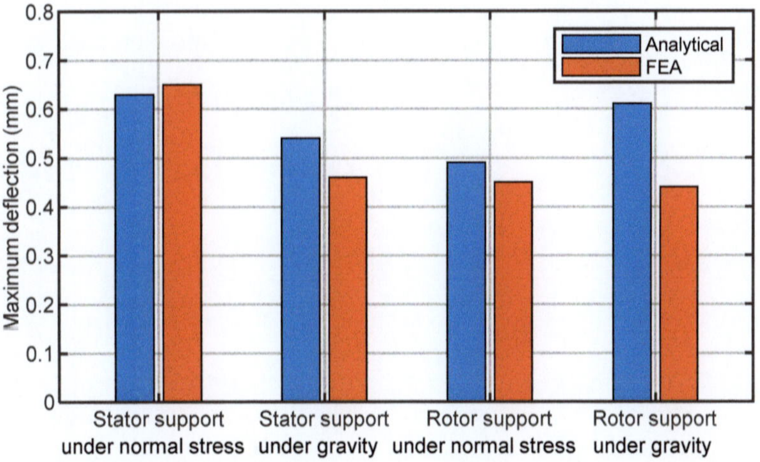

Fig. 4.5: The maximum deflection calculated by FEA and analytical model for stator and rotor support.

5 Design of a superconducting DC generator demonstrator

To pave the way for the superconducting generators from laboratory to industry, associated key technologies needs to be demonstrated, such as fabrication and operation of superconducting windings, protection of the superconducting coils in case of quench, availability of the excitation system for the superconducting coils, validation and reliability of the cryogenic system, robustness of the torque transfer system and so on. A common way to demonstrate these key technologies is to design, manufacture and test a downscaled demonstrator instead of the fullscale generator. For example in Superapower projectt, a downscaled prototype with two superconducting field coils was designed and built for a 10 MW design [64]. Through a downscaled prototype, the feasibility of the conceptual design can be validated at a lower effort. Moreover, during this process, major challenges to build a superconducting generator shall be learned and experience to solve the problems shall be gained. Due to the aforementioned advantages and to save the budget and efforts, a 10 kW superconducting DC generator demonstrator is designed.

5.1 Specifications

The summary of the demonstrator specifications is illustrated in Table 5.1. The rated voltage of the demonstrator is selected to be 230 V in DC, which is equal to the rectified voltage of the phase to ground voltage 400 V in AC throughout Germany. The insulation class of the copper winding is Class 155 and forced air cooling is used. The average wind speed is assumed to be 12.47 m/s, cut-in speed is 3.2 m/s, and cut-off speed is 24.6 m/s according to [143]. In order to transfer the wind speed into the generator rotational speed, eq. (5.1) and (5.2) will be used. The air density is assumed to be 1.225 kg/m^3. Applying the two aforementioned equations, we can get the power curves with rotational speed, as shown in Fig. 5.1. It can be seen that at the same wind speed with different rotational speeds, the mechanical power is different. As the output power of the generator is 10 kW and the assumed rated efficiency is around 85%, the mechanical power is 11.8 kW. In order to achieve the maximum mechanical power of 11.8 kW at wind speed 12.47 m/s, the rotational speed shall be 389 (1/min), at which point the rotor radius of the wind turbine is 2.57 m and the pitch angle is 0°. As the wind speed changes all the time, it is important to know the performance of the generator at different wind speeds. The turbine is assumed to be controlled by pitch control concept. At low wind speeds, the turbine is operated at maximum aerodynamic efficiency up to a rated speed of 389 (1/min); while at high speeds, the pitch angle is adjusted to maintain a rated speed of 389 (1/min).

Table 5.1: Design specifications of the demonstrator

Parameter	Value	Unit
Rated wind speed	12.47	m/s
Cut-in wind speed	3.2	m/s
Cut-off wind speed	24.6	m/s
Turbine rotor radius	2.57	m
Rated power	10	kW
Rated speed	389	1/min
Rated voltage	230	V
Copper winding insulation class	155	--
Copper winding cooling method	forced air cooling	--
Working temperature of superconductors	30	K

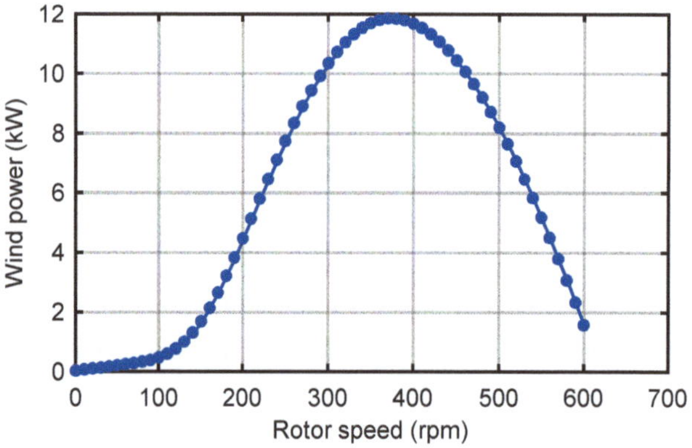

Fig. 5.1: Mechanical power cure with rotational speed.

$$P = \frac{1}{2}\rho_{air} C_p(\lambda, \theta_p)\pi R^2 v_w^3 \tag{5.1}$$

$$\begin{cases} \lambda = Rw_m / v_m \\ \xi = \dfrac{1}{\lambda + 0.08\theta_p} - \dfrac{0.035}{\theta_p^3 + 1} \\ C_p(\lambda, \theta_p) = 0.5176(116\xi - 0.4\theta_p - 5)e^{-21\xi} + 0.0068\lambda \end{cases} \tag{5.2}$$

where, P is the mechanical power, ρ_{air} is air density, R is the radius of the rotor, v_w is the wind velocity, C_p is the rotor power coefficient, which is the function of tip speed ratioλ and pitch angle θ_p.

5.2 Electromagnetic design

5.2.1 Critical current model of superconducting tapes

Superconducting tapes from four different companies, Sunam, Superox, Theva and Superpower were investigated for the demonstrator.The demonstrator and the superconducting coils are designed tooperate at 30 K with a conduction cooling to generate the main magnetic field in a 10 kW DC generator. Based on the data provided by the companies, the critical current as a function of the magnetic field was simulated. The Kim model is employed for the Superox and Theva data, which is shown in eq. (5.3) [144],[145]. In this part, only the perpendicular field is considered.

$$L(B,T)=\frac{J_C}{J_C(77K,self)}=\frac{j_{C0}(T)}{1+\left(\frac{B}{B_0(T)}\right)^{\beta}} \tag{5.3}$$

Where B is the perpendicular field applied on the tape, T is the tape temperature, $L(B,T)$ is the lift factor with applied magnetic field B and temperature T, J_c is the critical current density with applied magnetic field B and temperature T, J_c(77K, self) is the critical current density at 77 K and only self field is considered, $j_{c0}(T)$, $B_0(T)$ are coefficients with temperature T and β is related to the junction geometry. The lift factor is the ratio of critical current at a certain magnetic field and temperature to the critical current at 77 K self field. The fitting parameters, β, j_{c0} and B_0, at 30 K for Superox tapes are 1.0, 6.57 and 0.84; while the fitting parameters for the Theva tapes are1.0, 4.4 and 2.0.

The lift factor calculation for Sunam and Superpower tape is according to eq. (5.4) [146], and the fitting parameters for Sunam tape with perpendicular field, L_0, B_{c0}, and b, are 4.58, 0.49, and 1.14, respectively. The fitting parameters for Superpower tape with perpendicular field, L_0, B_{c0}, and b, are 6.12, 3.23, and 1.41.

$$L(\boldsymbol{B},T)=\frac{J_c(\boldsymbol{B},T)}{J_c(77\ K,self)}=\frac{L_0(T)}{\left(1+\sqrt{k(T)^2 B_x^2+B_y^2}\Big/B_{c0}(T)\right)^{b(T)}} \tag{5.4}$$

where $\boldsymbol{B}$ is the field applied to the superconducting tapes, which consists of parallel field component B_x and perpendicular field component of B_y; $k(T)$, $L_0(T)$, $B_{c0}(T)$, and $b(T)$ are coefficients with temperature T.

Based on the equations above, the lift factor of critical current of the tapes from the four companies as a function of magnetic fields is presented in Fig. 5.2. The critical current of the tapes at 77 K, self-field, and tape dimensions are listed in Table 5.2.

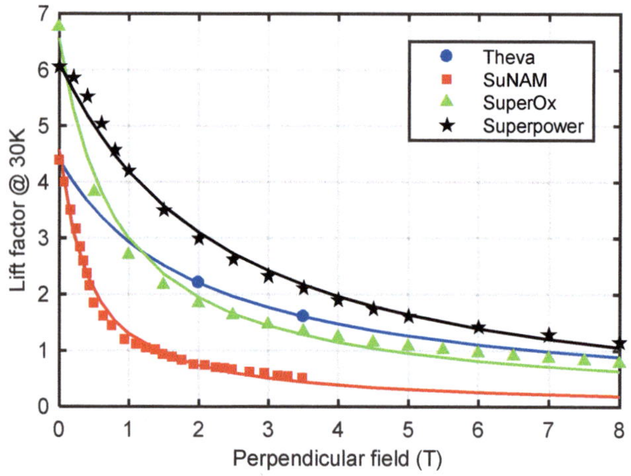

Fig. 5.2: Simulation of lift factor at 30 K for different tapes.

Table 5.2: Critical current, and dimensions of tapes

Parameter	SuNAM	SuperOx1	SuperOx2	SuperOx3	Theva	Super-power
Ic(77 K, self field) (A)	150	100	120	140	100	140
Thickness (mm)	0.23[a]	0.15	0.15	0.15	0.225[a]	0.164
Width (mm)	4	4	4	4	4	4

a. If the company only offers the tape without insulation, 0.0125 mm polyamide with 0.0125 mm silicone adhesive and overlap is assumed for insulation.

5.2.2 Comparison of designs with different tapes

By applying data in Table 5.2 and Fig. 5.2 into the analytical design model of the 10 kW superconducting generator described in Chapter 2, major design parameters as a function of average airgap flux density are shown in Fig. 5.3. The design constraints are: the length of tape used in a superconducting coil is less than 1 km, the armature diameter is 250 mm, the pole pair number is from 2 to 4, and the rotation speed of the generator is 389 (1/min). The flux density of *x*-axis in Fig. 5.3 the average flux density in the middle of the armature winding. From Fig. 5.3 it can be seen that the lowest weight and volume can be reached around 0.4 T, while the efficiency can be more than 90%. The tapes from different vendors almost have no influence on efficiency, as no AC losses are considered.

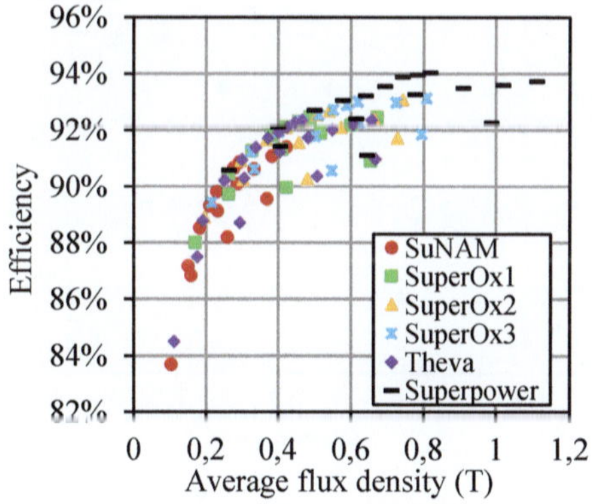

(a)

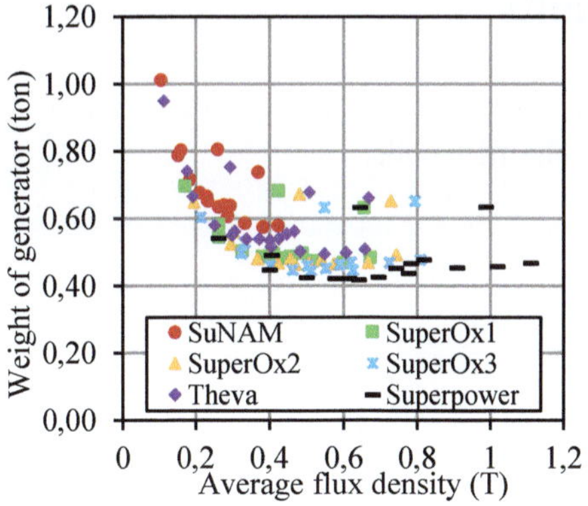

(b)

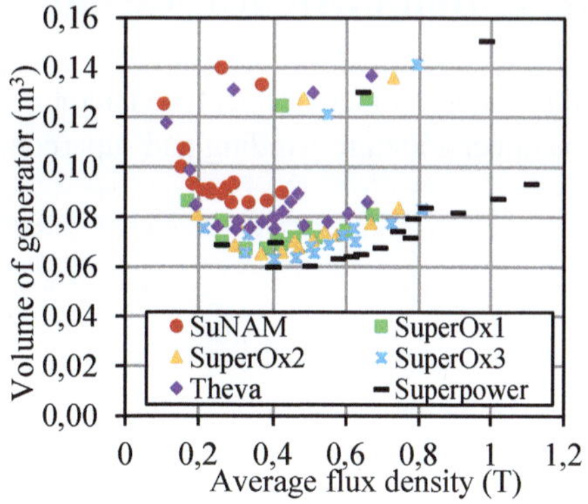

(c)

Fig. 5.3: Major design parameters as a function of the average air gap flux density of the 10 kW generator with different tapes, (a) efficiency of the generator (refrigeration losses is not included), (b) weight of the generator, and (c)overall volume of the generator.

The influence of the pole pair number on the major design parameters of the demonstrator is illustrated as Fig. 5.4. The influence of the pole pair number has the same trend for the 6 tapes listed in Table 5.2. As an example, tapes from Theva are used and the results are highlighted in Fig. 5.4. To provide the same average flux density in the armature winding a pole pair number of three has been chosen because of its lowest weight and volume.

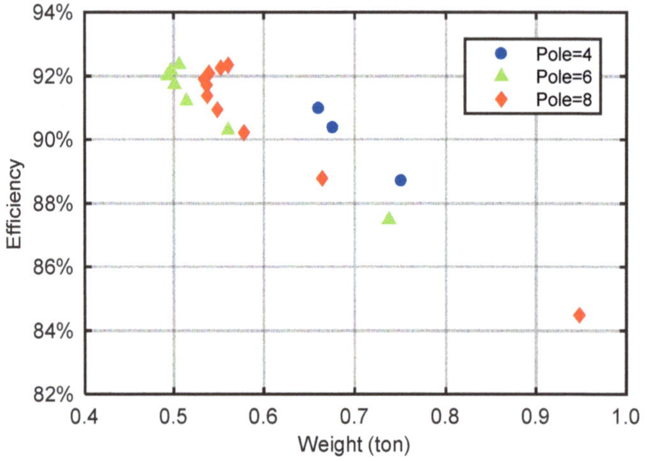

Fig. 5.4: Influence of the pole pair number on the major design parameters of the generator.

5.2.3 Demonstrator design results

A three-dimensional view of the generator is shown in Fig. 5.5. The generator employs air gap copper armature winding and superconducting field coils.

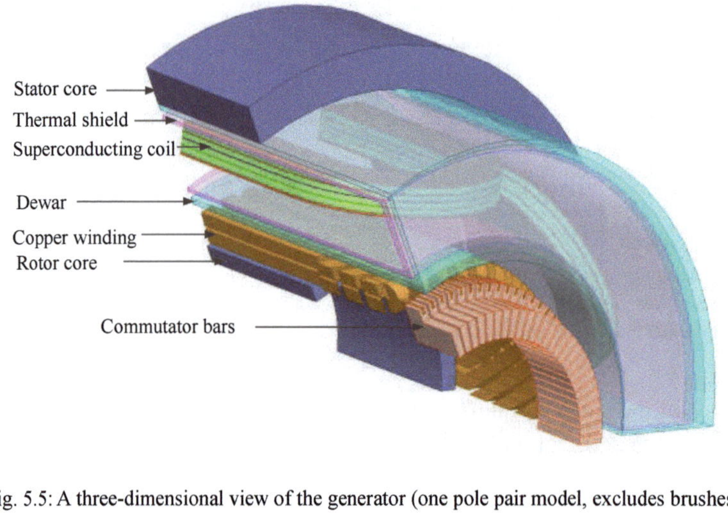

Fig. 5.5: A three-dimensional view of the generator (one pole pair model, excludes brushes).

The maximum perpendicular field in the superconducting coil is 1.24 T, which decides the operating current of the coil. The Kim model of the lift factor is used to calculate the critical current as shown in eq. (5.3). Hence, a lift factor of 2.72 at this magnetic field can be assumed. With a safety factor of 0.54, the operating current is selected to be 147 A. The design parameters and specifications of the generator are listed in Table 5.3.

Table 5.3: Design specifications and parameters of the 10 kW generator

Parameter	Value	Unit
Rated power	10	kW
Rated speed	389	1/min
Rated voltage	230	V
Number of poles	6	-
Number of stator coils	78	-
Stator outer diameter	486	mm
Stator inner diameter	260	mm
Stator core thickness	30	mm
Physical air gap	5	mm
Rotor outer diameter	250	mm
Rotor inner diameter	164	mm
Rotor core thickness	13	mm
Stack length	220	mm
Electrical loading	303	A/cm
Average flux density	0.52	T
Number of turns per superconducting coil	604	-
Number of turns per copper coil	21	-
Operating current	147	A
Superconducting tapes	3.2	km
Thickness of dewar wall	5	mm
Thickness of thermal shield wall	3	mm
Thickness of vacuum insulation	5	mm
Brush length	20	mm
Brush width	16	mm
Brush holders	6	--
Brush per brush holder	1	--
Commutator length	45	mm
commutator diameter	160	mm
Brush shift	3.2	°

A 3D magnetostatic model is used to calculate the flux density in the airgap winding and results are illustrated in Fig. 5.6. Radial flux density in different axial positions is highlighted in Fig. 5.6. It can be seen that along the stack length, the flux density decreases only slightly. Hence, the flux density in the middle axial position can be used to represent that along the stack length.Then we can simplify the calculation region from 3D to 2D surface. Hence, 2D FE model is used to simulate the generator. To analyse the generator performance, the FE model is coupled with a commutation drive model and the rated load in the simulation is considered as a DC current source with rated current 43.5 A. The calculated torque and branch voltage is presented in Fig. 5.7. The peak to peak ripple of the torque and voltage is 1.26% and 3.26%, which can be decreased by employing more coils with less turns. The coil current and induced voltage is demonstrated in Fig. 5.8. To limit the voltage between two commutator bars not to exceed a maximum value, typically 30 to 40 V, 78 commutator bars are used, as a result from eq. (5.5).

$$N_{com} \geq \frac{2pU_N}{V_{c\max}} \tag{5.5}$$

where N_{com} is the number of commutator segments, V_{cmax} is the maximum value of he voltage between two commutator segments.

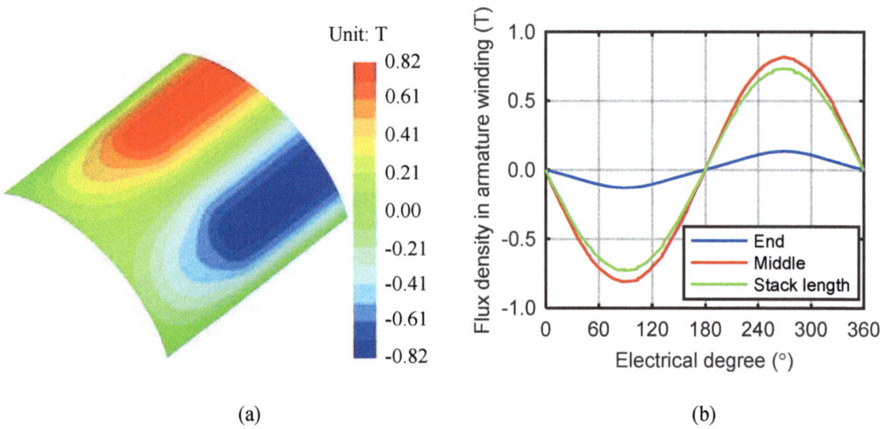

Fig. 5.6: Radial component of flux density in the armature winding: (a) flux density distribution in the middle radial position, (b) flux density at different axial positions.

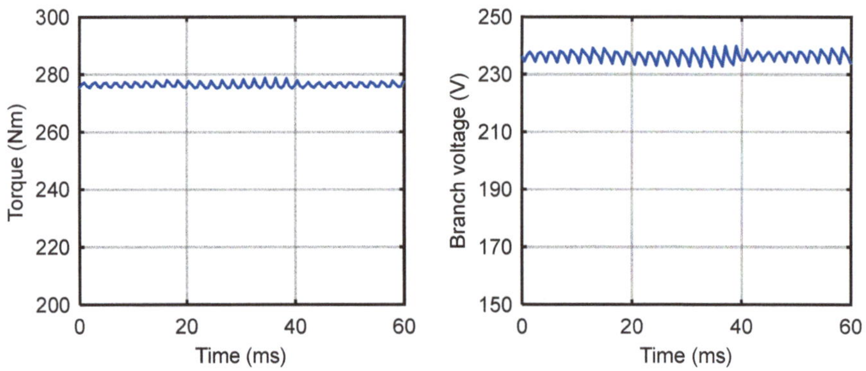

Fig. 5.7: Torque and branch voltage of the generator.

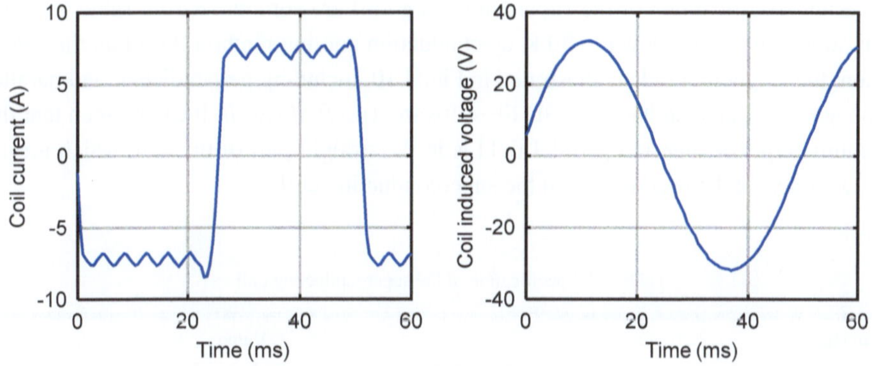

Fig. 5.8: Coil current and voltage of the generator.

5.3 Design of the superconducting coils

The structure of the superconducting coil that consists of two double pancakes (DP) is shown in Fig. 5.9. The coil has 2 double pancakes and there is an insulation layer, made of G10, between the pancakes. Within a double pancake, a copper layer is inserted between the two layers of the superconducting tapes for cooling, as well as the two copper caps on the top and bottom of the coil.

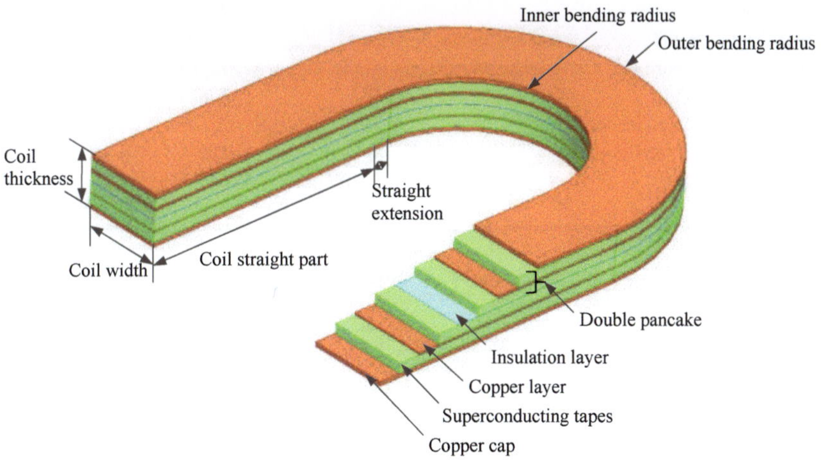

Fig. 5.9: Structure of the superconducting coil.

The design specifications of the superconducting coil are summarized in Table 5.4. The coil is supposed to cool down to 30 K by conduction cooling method.Based on the specification, the magnetic field, as illustrated in Fig. 5.10, including perpendicular and parallel component, is calculated based on 3D FE software *Ansoft Maxwell*. It can be seen that the maximum perpendicular and parallel field is in the straight part of the coil, which means that 2D model can be used to design the superconducting coil.

Table 5.4: Specification of the superconducting coil

Parameter	Value
Number of double pancakes (DPs)	2
Total thickness of the coil	22.8 mm
Total width of the coil	36.0 mm
Number of turns of the coil	604
Number of copper caps	2
Thickness of copper caps	2 mm
Insulation layer between DPs	0.2 mm/G10
Coil straight part length	220 mm
Coil straight extension	10 mm
Coil bending radius (inner/outer)	45/81 mm
Insulated tape size (mean)	4.15 mm ×0.225 mm
Copper layer within a DP	1mm
Thickness of a DP	9.3 mm
Width of a DP	36.0 mm
Terminal length	100 mm
Tape length per DP	266 m
Total tape length per coil	532 m
Maximum perpendicular field	1.24 T
Maximum parallel field	1.93 T
Lift factor @Max. perpendicular field, 30K	2.716
Critical current @77K, self-field	100 A
Operating current	147 A
Inductance per coil	0.22 H

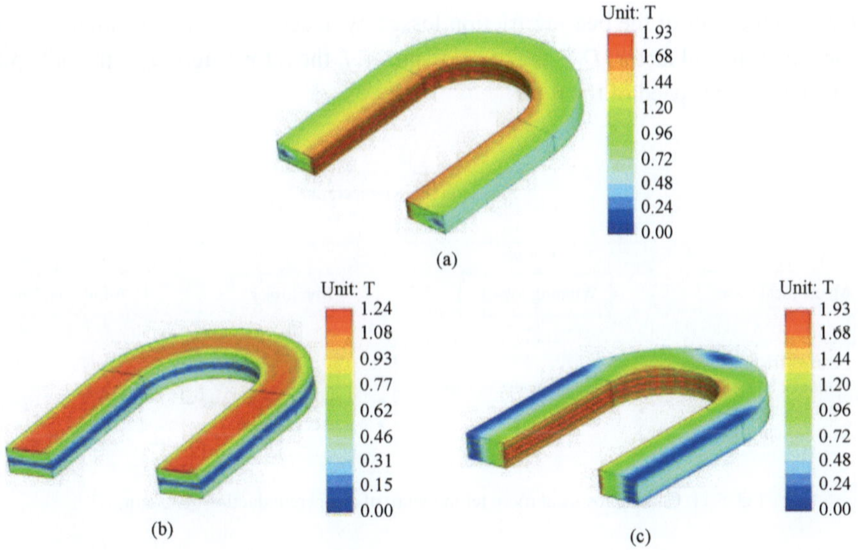

Fig. 5.10: Distribution of the magnetic field,(a) magnitude of the flux density,(b) magnitude of perpendicular component, and (c) magnitude of parallel component.

5.4 Losses in the demonstrator

5.4.1 Losses at room temperature

The generator losses at room temperature are illustrated in Fig. 5.11. In general, in the DC superconducting generator, there are mechanical losses P_m, winding losses P_w, iron losses P_{Fe} and additional losses P_{add} at room temperature. Winding losses consist of copper resistive loss P_{Cu}, and brush contact loss P_b. Additional losses are losses in windings, laminations, frame and other construction parts due to load current and its spatial harmonics, which are not included in the copper resistive losses and iron losses. Winding losses and additional losses can be calculated by eq (2.18), (2.19), (2.20) and (2.21). Mechanical losses are a consequence of bearing friction, windage, ventilator and brush friction losses.

The empirical equation to calculate the bearing friction losses, windage losses, and ventilator losses is expressed as [33]:

$$P_{m1} = k_{m1} D_r \left(l_r + 0.6\tau_p \right) v_r^2 \tag{5.6}$$

where P_{m1} is the sum of the bearing friction losses, windage losses, and ventilator losses; k_{m1} is an experimental factor; D_r is the rotor diameter, l_r the rotor length, τ_p is the pole pitch and v_r is the surface speed of the rotor.

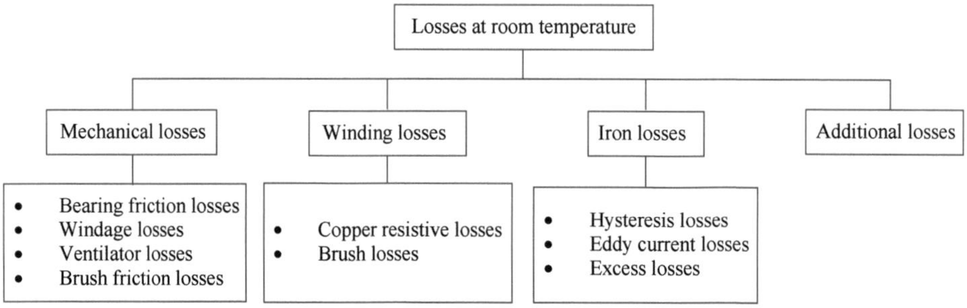

Fig. 5.11: General losses at room temperature of a superconducting generator.

The brush friction losses can be estimated by

$$P_{m2} = k_{m2} P_{rB} A_{rB} v_{rB} \tag{5.7}$$

where P_{m2} is the brush friction losses; k_{m2} is an friction coefficient; P_{rB} is the brush average pressure density (N/m^2); and v_{rB} is the surface speed of the brush. As a result, the losses in the room temperature are summarized in Table 5.5.

Table 5.5: Summary of losses at room temperature

Parameter	Value	Equation
Copper resistive losses	698.7 W	Eq. (2.18)
Iron losses	23.2 W	Eq. (2.19)
Additional losses	100.0 W	Eq. (2.20)
Brush contact losses	113.1 W	Eq. (2.21)
Sum of bearing, windage, and ventilator losses	32.1 W	Eq. (5.6)
Brush friction losses	24.6 W	Eq. (5.7)
Total losses	1008.6 W	

Axial forced air cooling is employed to cool the copper winding in the rotor and the rotor cooling design is illustrated in Fig. 5.12. To estimate the maximum temperature in the copper winding, a steady-state thermal analysis is performed and result shows that the maximum temperature in the copper winding and the winding support, which is 134.5 °C. This is within the limit of F-class insulation (155 °C).

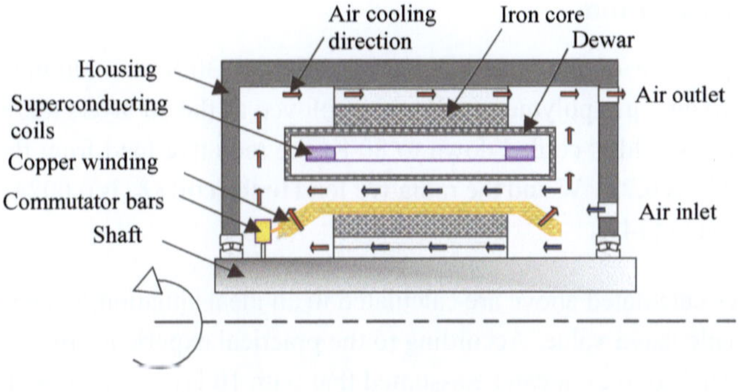

Fig. 5.12: Sketch of axial forced air cooling.

5.4.2 Losses in at cold temperature

The structure of the cryostat and all the heat loads at cold temperature are clarified in Fig. 5.13. Even though the superconducting coils operate with DC current, there are still AC losses due to the ramping current at start-up and ripples from current supply, which are denoted as P_{ac1} and P_{ac2}. Copper current leads are employed to feed current into the superconducting coils from the current supply. Unavoidly, copper current leads will introduce heat, P_{cl1} and P_{cl2}. To support the superconducting coils and thermal shield, support rods are used. Hence, heat loads due to support rods, P_{rod1} and P_{rod2}, needs to be removed. Moreover, instrument wires, which is used to measure the temperature, magnetic field, and superconducting coils voltage, *et al* in the cryostat also cause a small portion of heat load, P_{mw}. Besides, there are radiative losses, P_{r1} and P_{r2}, from the room to the cold temperature.

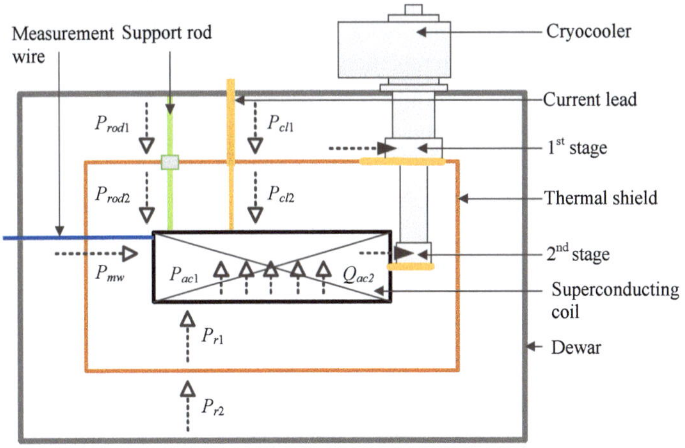

Fig. 5.13: Heat loads at cold temperature.

5.4.2.1 Thermal radiation

The working temperature of the superconducting tapes is 30 K. If aluminium coating (79 nm in thickness) on 6 μm polymester film is employed as the 10 layers thermal insulation, and the thermal shield is cooled down to 80 K, the radiative load from the dewar to the thermal shield Q_{r1} is 0.721 W and the radiative load to the coil Q_{r2} is 0.0024 W according to the calculation in [147].

The radiative losses calculated above are calculated in an ideal situation, in fact, the losses is higher than the calculated value. According to the practical experience of cryogenic experts in ITEP (internal source), it can be assumed that with 10 layers of thermal insulation the heat flux density is 2 W/m^2 from room temperature to 77 K. As the total surface area of the dewar is about 0.96 m^2, the radiative loss to the thermal shield Q_{r1} can be estimated to be 1.92 W when the thermal shield is cooled down to 80 K, while the radiative load from the thermal shield to the coil is much less than that since the temperature is from 80 K to 30 K. As an estimation, 10 % of 2 W/m^2 is used. As a result, the radiative load from 80 K shield to the superconducting coil at 30 K Q_{r2} is 0.19 W.

5.4.2.2 AC losses due to the superconductors

Although the current of the superconducting coils is designed to be DC, ac magnetic fields will act on the superconductors from start up, dc current source ripple, ripples from the armature windings, etc. As the effective airgap is very large in our design, the influence from the armature winding is quite small. The magnetic field ripples caused by the armature winding are in the order of 0.07 mT with a frequency of 19.45 Hz or 252.85 Hz, while the magnetic field of the superconducting coils is 1.94 T. Compared the magnetic field ripples to the magnetic field, they are far too small. Hence, the influence of the armature winding current in normal operation is neglected. In this section, only the ramping current and ripples of the operating current are considered. In the superconductor, the resistivity is not constant when the magnetic field, temperature and flowing current changes. To describe the macroscopic behaviour of it, a nonlinear E-J relation is employed as shown in eq (5.8). Since $E = \rho_{sc} J$ then the resistivity of the superconductor, ρ_{sc}, is expressed in eq (5.9).

$$E = E_c \left(\frac{J}{J_c}\right)^n \tag{5.8}$$

$$\rho_{sc} = \frac{E_c}{J_c} \left(\frac{J}{J_c}\right)^{n-1} \tag{5.9}$$

where, E_c corresponds to the critical electric field , which is 10^{-4} V/m, J_c is the critical current, n is the n-value indicating the sharpness of the E-J characteristic, which is 38 in this section [148]-[150]. The critical current has a dependence of magnetic field at 30 K, which is expressed as eq (5.2).

To calculate the AC losses, *H*-formulation , as in eq (5.10), is used.

$$\nabla \times \rho \nabla \times H = -\mu \frac{\partial H}{\partial t} \tag{5.10}$$

where, ρ is the resistivity of the material, μ is the permeability of the material, and H is the magnetic strength.

To increase the computation speed, the homogenization and multiscale technique are adopted [148]-[150]. With this method, the superconducting coils are regarded as an anisotropic bulk and the actual different layers of the superconductors are not considered. In fact, the actual topological features of the tapes are considered by a superconducting material's volume fraction to calculate the engineering current density. A multiscale model is employed by treating only one side of a superconducting coil with superconducting characteristics, while the others are treated as "normal conductors", which has a constant resistivity and the current density is uniform distributed. In this way, the magnetic flux density distribution around the homogenized model is almost the same as in the generator and no extra boundaries are needed.

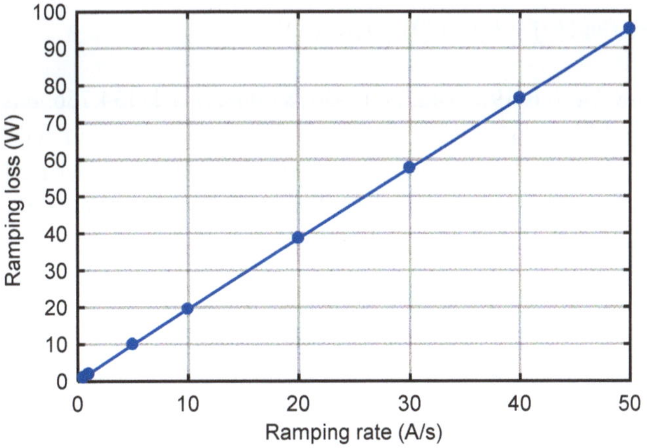

Fig. 5.14: Ramping loss with different ramp rates.

To calculate the ramping current loss, a smooth ramp up current is used to increase to the operating current, which is 147 A. the results is illustrated in Fig. 5.14. Different ramping speed contributes to different ramping current losses, and the lower the ramping current rate, the smaller the losses. Hence, a ramping rate below 0.5 A/s is preferred to have less loss. The ramping current loss Q_{ac1} is below 1.01 W.

The current ripples are assumed to be at 0.8 Hz with amplitude of 1% of the operating current [61]. The resulting ripple current loss Q_{ac2} is 3.70 W. When the current ripple increased to 2% with a frequency of 2 Hz, the calculated loss will be increased to 5.60 W.

5.4.2.3 Losses due to support rods

The material used for the torque tubes is titanium alloy, whose maximum yield strength is 1132 MPa at 4-40 K [151]. The torque tube has to withstand the electromagnetic torque, as well as to minimize the heat load to the cold mass. In this part, tie rods made of Ti6Al4V are employed to suspend the coils in the Dewar, similar to the mechanical structure in a magnetic resonance imaging (MRI) [152]. There are radial tie rods and axial tie rods. The radial and axial tie rods are shown in Fig. 5.15 (a), and Fig. 5.15 (b). Moreover, a copper braid from the thermal shield is adhered to the surface of a rod where it crosses the thermal shield. Hence, there is a temperature breakpoint, that is, the rod is divided into temperature range from 80 K to 300 K, and range from 30 K to 80 K. The main parameters of the support structure are listed in Table 5.6.

Thermal conductivity, k, of the titanium alloy is calculated by (5.11) [153].

$$\begin{aligned}\log(k) = a + b\log(T) + c(\log(T))^2 + d(\log(T))^3 + e(\log(T))^4 + f(\log(T))^5 \\ + g(\log(T))^6 + h(\log(T))^7 + i(\log(T))^8\end{aligned} \tag{5.11}$$

where, a is -5107.8774, b is 19240.422, c is -30789.064, d is 27134.756, e is -14226.379, f is 4438.2154, g is -763.07767, h is 55.796592, and i is 0, respectively; T is the temperature in K. As a result, the thermal conductivity of the titanium alloy is underlined in Fig. 5.16. The average thermal conductivity between 80 K and 300 K is 5.464 W/(m•K), and the average value between 30 K and 80 K is 2.621 W/(m•K).

Table 5.6: Parameters of rods

Name	Radial rods	Axial rods
Number of rods	24	6
Diameter of rods	5 mm	5 mm
Length (80-300 K)	36 mm	178 mm
Length (30-80 K)	80 mm	179 mm

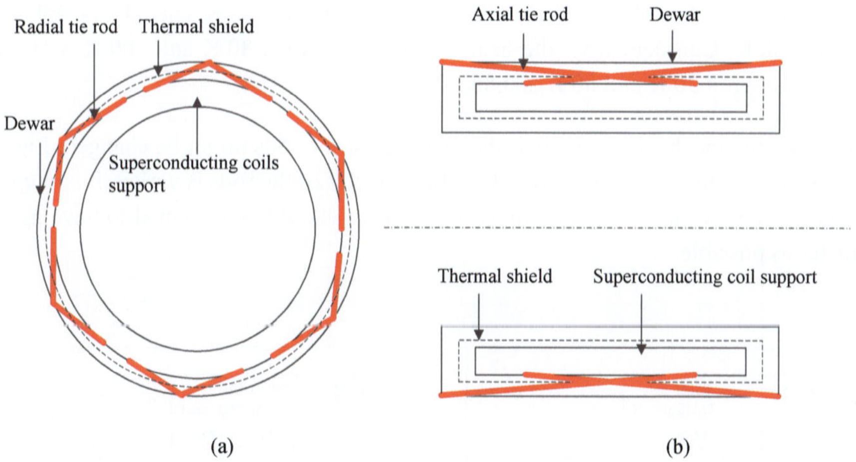

Fig. 5.15: Illustration of superconducting coil support, (a) radial rods, (b) axial rods.

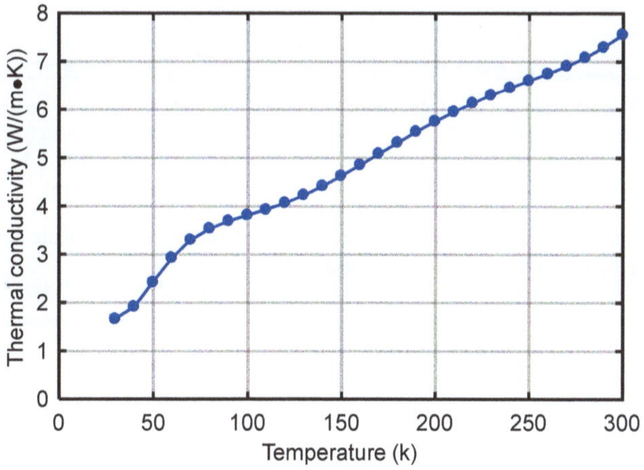

Fig. 5.16: Calculated thermal conductivity of the titanium alloy Ti6Al4V.

Then we can calculate the Heat leakage Q_{rod} of titanium alloy by (5.12)[154]

$$Q_{rod} = \int_{T1}^{T2} \frac{A_{rod}}{L_{rod}} \lambda_{ti}(T)dT \tag{5.12}$$

where, L_{rod} is the length of the support rod, A_{rod} is the cross section per rod, T is temperature, λ_{ti} is thermal conductivity of titanium alloy, T_1, T_2 is the temperature of the two ends of the rod. By this method, the conductive heat leak of a single radial rod is calculated to be 0.6556 W between 80 K and 300 K, and 0.0322 W between 30 K and 80 K. The conductive heat leak of a single axial rod is 0.1326 W between 80 K and 300 K, and 0.0144 W between 30 K and 80 K. Consequently, the heat load Q_{rod1} between 80 K and 300 K is 16.54 W, and Q_{rod2} between 30 K and 80 K is 0.86 W.

To increase the mechanical stability, the diameter of the rods might be changed. Fig. 5.17 pictures the heat load of the tie rods when the diameter of the rods is varied. It is suggested that within the mechanical requirement, the tie rods should be designed to have as small diameter as possible.

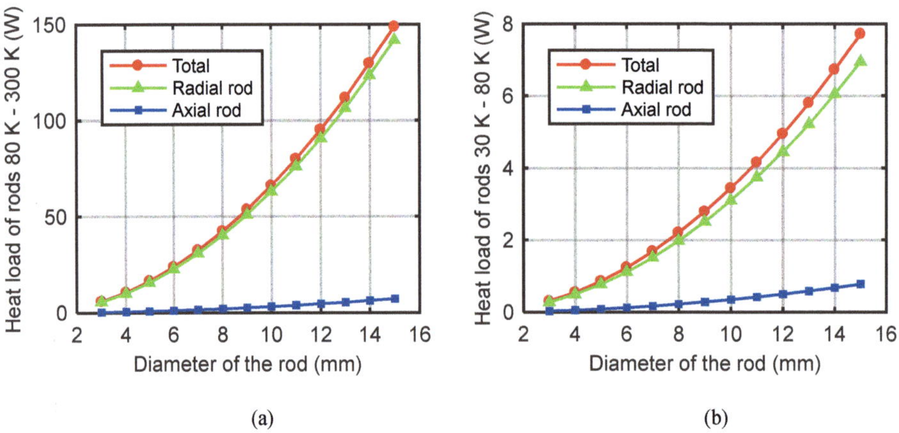

Fig. 5.17: Heat load of the tie rods when the diameter of the rods is varied,(a) temperature between 80 K and 300 K, and (b) temperature between 30 K and 80 K.

5.4.2.4 Losses due to current leads

The electrical resistivity of the copper ρ_{cu} depends on the temperature and RRR value, which can be obtained through eq. (5.13) [155].

$$\rho_{cu}(T,\mathrm{RRR}) = \rho_0 + \rho_i + \rho_{i0} \tag{5.13}$$

where:

$$\rho_0 = \frac{1.553\times10^{-8}}{\mathrm{RRR}} \tag{5.14}$$

$$\rho_i = \frac{P_1 T^{P_2}}{1 + P_1 P_3 T^{P_2 - P_4} \exp\left[-\left(\frac{P_5}{T}\right)^{P_6}\right]} \tag{5.15}$$

$$\rho_{i0} = P_7 \frac{\rho_i \rho_0}{\rho_i + \rho_0} \tag{5.16}$$

The magetoresitivity effect is considered by adding magnetic field, B, to eq. (5.17).

$$\rho_{cu}(T, \mathrm{RRR}, B) = \rho_{cu}(T, \mathrm{RRR})\left(1 + 10^{a(x)}\right) \tag{5.17}$$

where:

$$a(x) = \sum_{n=0}^{4} a_n (\log_{10} x)^n \tag{5.18}$$

$$x = S(T, \mathrm{RRR}) \times B = \frac{\rho_{cu}(T = 273\mathrm{K}, \mathrm{RRR})}{\rho_{cu}(T, \mathrm{RRR})} \times B \cong \frac{1.553 \times 10^{-8}}{\rho_{cu}(T, \mathrm{RRR})} \times B \tag{5.19}$$

The fit parameters used in eq. (5.15), (5.16), and (5.18) are listed in Table 5.7.

Table 5.7: Fit parameters for copper electrical resistivity

P_1	P_2	P_3	P_4	P_5	P_6	P_7
$1.171 \cdot 10^{-17}$	4.49	$3.841 \cdot 10^{+10}$	1.14	50	6.428	0.4531

a_0	a_1	a_2	a_3	a_4
-2.662	0.3168	0.6229	-0.1839	0.01827

Note: there is a minimal value of B at which equation (5.17) fails. The minimal B at room temperature is in the order of 1 T, and it decreases with lower temperatures depending on RRR.

The analytical way to obtain the thermal conductivity of the copper is introduced by equation (5.20).

$$k_{cu}(T, \mathrm{RRR}) = \frac{1}{W_0 + W_i + W_{i0}} \tag{5.20}$$

where:

$$W_0 = \frac{\beta}{T} \tag{5.21}$$

$$W_i = \frac{P_1 T^{P_2}}{1 + P_1 P_3 T^{P_2+P_4} \exp\left[-\left(\frac{P_5}{T}\right)^{P_6}\right]} \tag{5.22}$$

$$W_{i0} = P_7 \frac{W_i W_0}{W_i + W_0} \tag{5.23}$$

The fit constants in eq. (5.21) (5.22) and (5.23) are reported in Table 5.8.

Table 5.8: Fit parameters for copper thermal conductivity

β	β_r	P_1	P_2	P_3	P_4	P_5	P_6	P_7
0.634/RRR	β/0.0003	1.754•10^{-8}	2.763	1102	-0.165	70	1.756	$0.838/\beta_r^{0.1661}$

As a result, the resistivity is increased with higher flux density. According to Wiedemann-Franz law:

$$\rho_{cu} k_{cu} = L_0 T \tag{5.24}$$

When the temperature is a fixed value, the electrical resistivity and thermal conductivity are inverse proportional to each other, where the L_0 is the Lorenz number, which is about 2.45×10^{-8} WΩ/K^2. Hence, the thermal conductivity is decreased with magnetic field. To calculate the resistivity of the copper current lead we assume a magnetic flux density of 2 T which is the maximum value in the superconducting coil, whereas a magnetic flux density of 0 T is assumed to calculate the thermal conductivity to have the worst case. Consequently, the calculated value is larger than the real value, which gives a larger safety margin. The electrical resistivity and thermal conductivity of copper with RRR=100 between 30 K and 300 K are highlighted in Fig. 5.18.

Heat leakage of the current lead Q_{cl} is

$$Q_{cl} = Q_J + Q_c = \int_{l_{l1}}^{l_{l2}} I_{sc}^2 \frac{\rho_{cu}}{A_l} dl + \int_{T1}^{T2} \frac{A_l}{l_l} k_{cu} dT \tag{5.25}$$

where I_{sc} is operating current of superconducting tapes, Q_J is ohmic losses, Q_c is conduction heat leakage, l_l is the length of the current lead, A_l is the cross section of the current leads, T is the temperature of the current lead, ρ_{cu} is the resistivity of copper, k_{cu} is the thermal conductivity, T_1 and T_2 are the temperature of the two ends of one current lead.

The calculated heat loss of one current lead by eq. (5.25) with different lengths and diameters is illustrated in Fig. 5.19. Due to the space limits in the generator, the length and diameter of the current lead between 30 K and 80 K are chosen to be 100 mm and 2 mm. Besides, the length and diameter of the current lead between 80 K and 300 K are chosen to be 250 mm and 4 mm. Hence, the heat load of two current leads, Q_{cl1} and Q_{cl2}, is 4.64 W between 30 K and 80 K and 18.21 W between 80 K and 300 K.

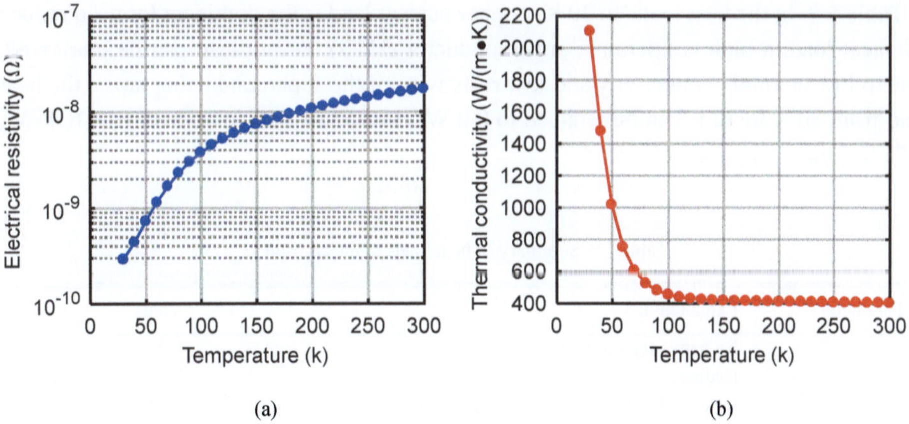

Fig. 5.18: The electrical resistivity and thermal conductivity of copper with RRR=100,(a) electrical resistivity, and (b) thermal conductivity.

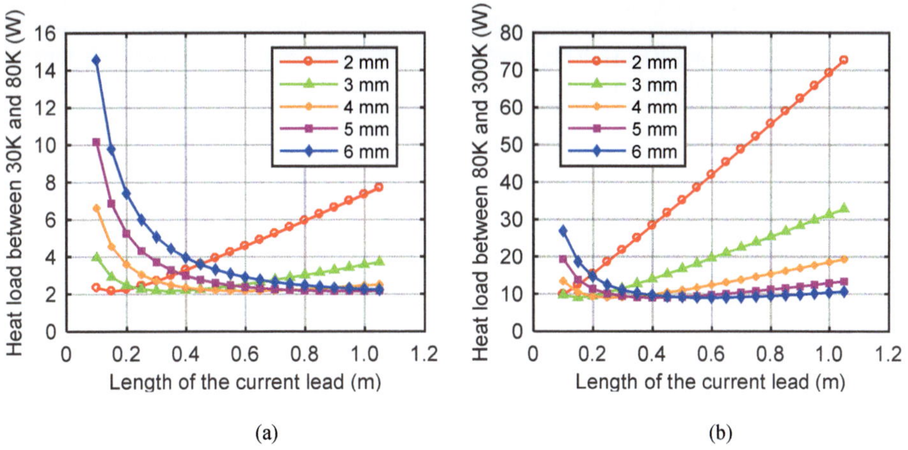

Fig. 5.19: Heat loss of one current lead with different lengths and diameters,(a) Heat load between 30 K and 80 K, and (b) heat load between 80 K and 300 K.

5.4.2.5 Losses due to measurement wires

As the measurement wires are very thin and usually mounted on a thermal anchor, the heat load due to measurement wires Q_m is very small and 0.1 W is estimated to take into account the measurement wires based on experience.

5.4.2.6 Summary of heat loads to the coild temperature

All the heat loads in the cryostat at low temperature as shown in Fig. 5.13 are summarized in Table 5.9. In the heat loads to 30 K, copper current lead is the dominant loss. To reduce this heat load, a high temperature superconductor (HTS) current lead can be employed. Due to low thermal conductivity and low resistivity of the superconducting tapes, the heat load from 80 K to 30 K can be reduced to 0.2 W. Hence, the heat load to 30 K is 6.06 W, 43% less than copper current lead.

Table 5.9: Summary of heat loads in the cryostat

Parameter	Explanation	Value	Unit
P_{r1}	Radiative load to 80 K	1.92	W
P_{r2}	Radiative load to 30 K	0.19	W
P_{rod1}	Heat load to 80 K caused by support rods	16.54	W
P_{rod2}	Heat load to 30 K caused by support rods	0.86	W
P_{cl1}	Heat load to 80 K caused by current leads	18.21	W
P_{cl2}	Heat load to 30 K caused by current leads	4.64	W
P_{ac1}	Ramping current loss	1.01	W
P_{ac2}	Current ripple loss	3.70	W
P_{mw}	Heat load from measurement wires	0.10	W
P_{s1}	Total heat load to 80 K	36.67	W
P_{s2}	Total heat load to 30 K	10.50	W

The efficiency of the refrigerators used in below 77 K is less than 20% of the Carnot efficiency [67]. The Carnot efficiency η_c can be calculated as

$$\eta_c = T_{low} / \left(T_{high} - T_{low}\right) \tag{5.26}$$

where, T_{low} is the low temperature side, T_{high} is the high temperature side.

If the ambient temperature is assumed to be 300 K, the refrigerator efficiency at 80 K is 7.27% and 2.22% at 30 K. The refrigeration power at room temperature to remove the heat in Table 5.9 is 977 W.

5.5 Performance at different wind speeds

Wind turbines normally operate with a range of wind velocity defined by maximum and minimum wind velocity. The turbines start to deliver power at the minimum wind velocity which is called cut-in velocity and the turbines stop operation at the maximum wind velocity which is called cut-out velocity. When the wind velocity is between the cut-in and rated wind velocity, the turbines are controlled to capture as much energy as possible with a fixed pitch angle, which is called the maximum power point tracking (MPPT). When the wind velocity is between the rated and cut-off velocity, the generators will be controlled with the same power output and rotational speed. In that case, the generator torque is fixed, and the pitch angle of the turbines is varied to scale down the aerodynamic torque [156],[157]. At different wind speeds, the generator voltage is proportional to the rotational speed and current is controlled according to the wind turbines. The wind turbine extracted wind power, rotational speed with wind velocities are shown in Fig. 5.20.

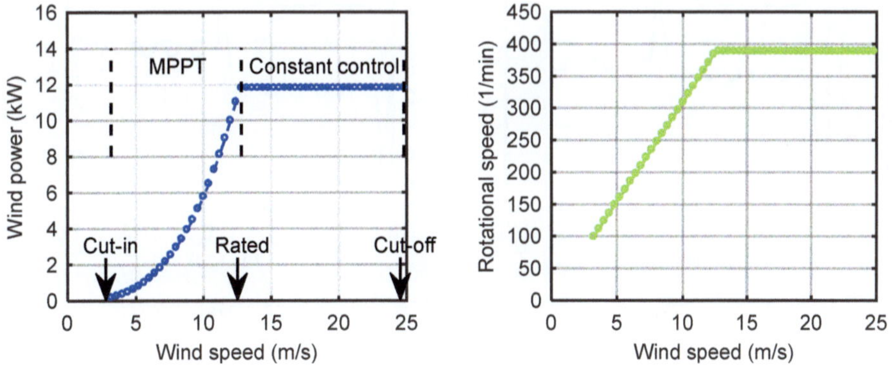

Fig. 5.20: Sketch of wind power and rotor speed with wind velocities.

5.5.1 Commutation

Armature coils undergoing commutation will induce an electromotive force (EMF) due to its inductance and current changes. Moreover, in the demonstrator the flux density distribution established by the superconducting coils is close to sinusoidal, and the commutation zone is penetrated by it. Due to the coil rotation, another EMF is induced in the coil undergoing commutation. Both EMFs will result in commutation challenges, for example, sparks. Compensation methods are required. In general, interpoles and compensation windings are employed to smooth commutation. However, in a superconducting DC generator, it is improper to adopt such methods, since the stator is excited by the superconductors and has no iron poles. Hence, to shift the brushes is used to improve commutation.

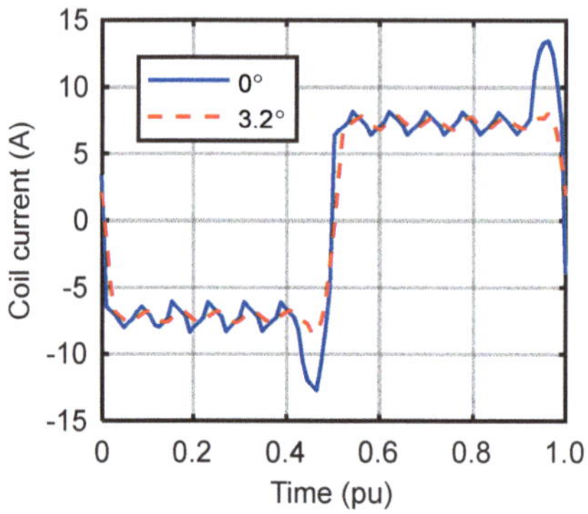

Fig. 5.21: Current per coil with/without brush shift at rated rotational speed.

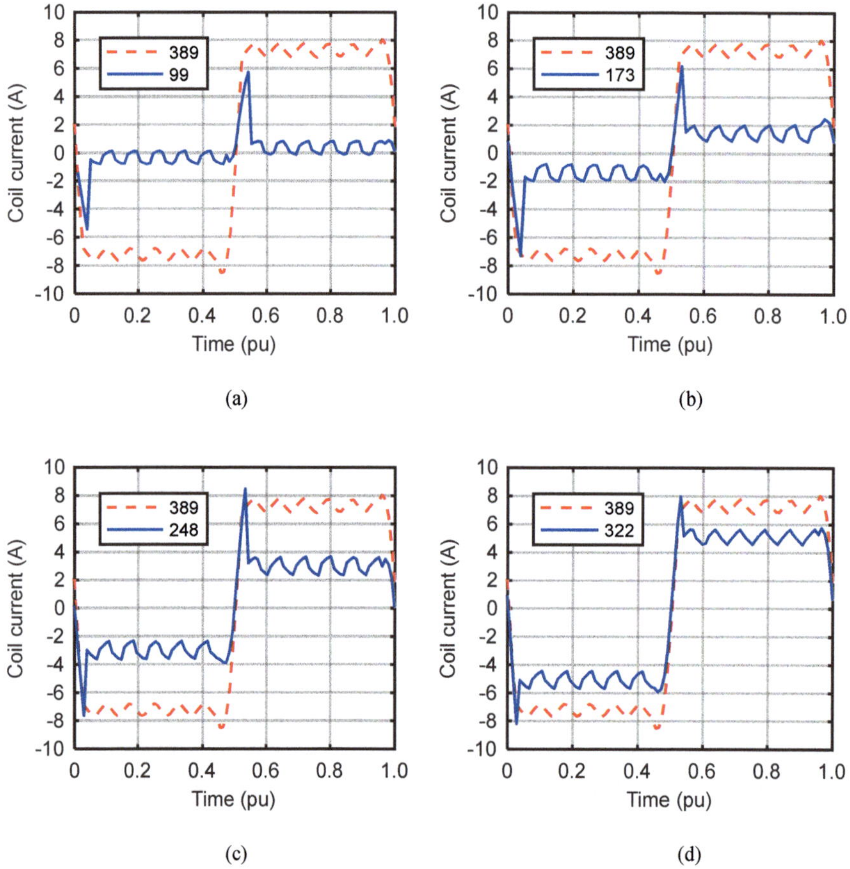

Fig. 5.22: Current per coil at different rotational speeds with brush shift 3.2°, (a) 99 (1/min), (b) 173 (1/min), (c) 248 (1/min), and (d) 322 (1/min).

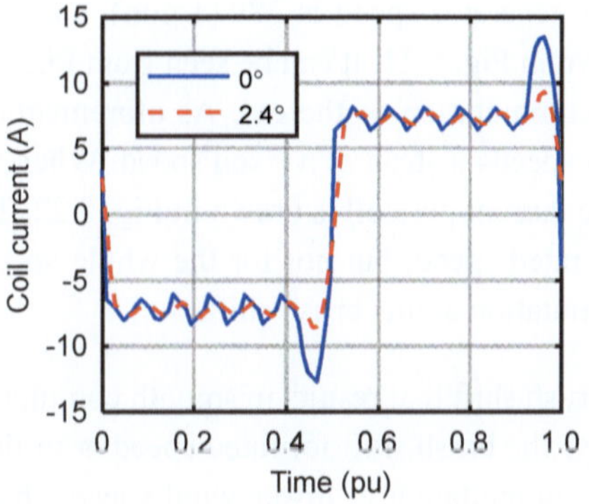

Fig. 5.23: Current per coil with brush shift 2.4° and without brush shift at rated rotational speed.

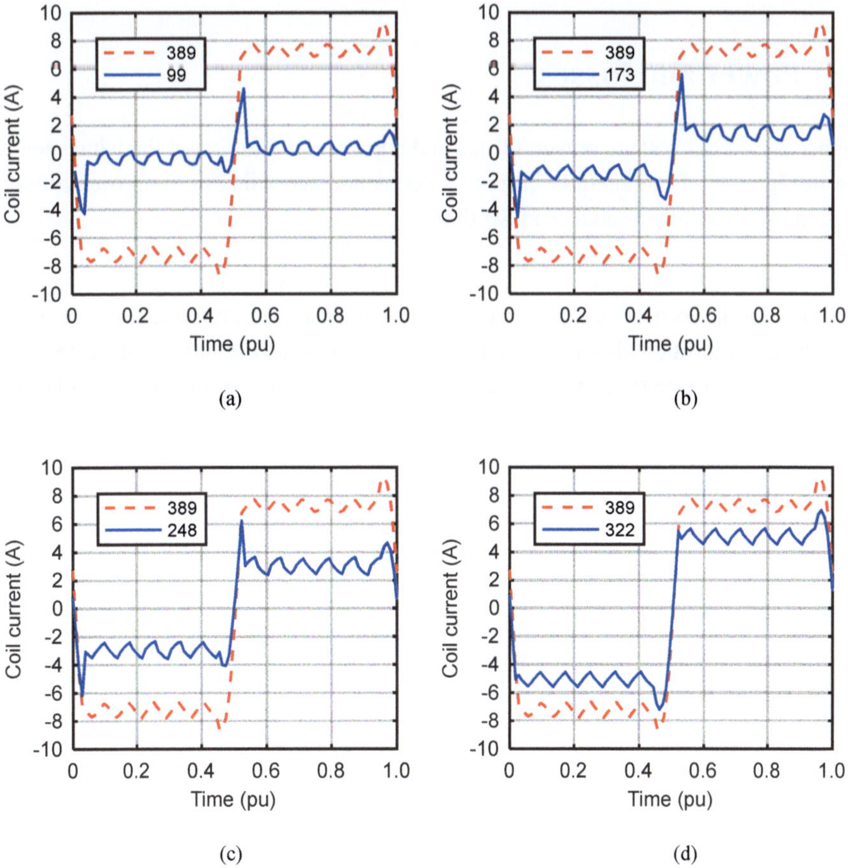

Fig. 5.24: Current per coil at different rotational speeds with brush shift 2.4°, (a) 99 (1/min), (b) 173 (1/min), (c) 248 (1/min), and (d) 322 (1/min).

At rated wind speeds, the generator speed is 389 (1/min), and the commutation with or without brush shift is shown in Fig. 5.21. It can be seen from Fig. 5.21 that a brush shift by 3.2° results in a smooth commutation of the coi. As aforementioned, the wind turbines operate at a range of wind speeds instead of a fixed speed. When wind turbines operate at differrent wind speeds, the current per coil is drawn in Fig. 5.22. The brush shift of 3.2° is good for commutation at rated speed, but not for the whole speed range. The lower the speed, the worse the commutation at this brush shift.

It is impossible to find a brush shift that results in smooth commutation at all wind speeds. The solution is either to use the brush shift for rated speed or to find a compromised solution, which improves the commutation at lower wind speeds, but decreases at the rated speed to some extent. When the brush shift is modified to 2.4°, the coil current at different wind speeds are shown in Fig. 5.23 and Fig. 5.24. For our demonstrator we mainly focused on the performance at rated speed, hence, a brush shift of 3.2° is used.

5.5.2 Power and losses

The generator output voltage power, torque and current as a function of the wind speed are shown in Fig. 5.25. The generator current is controlled according to the power converted through the wind turbine model as in Section 5.

Losses and efficiency with the wind velocities are illustrated in Fig. 5.26. It should be pointed out that the refrigeration power at room temperature to remove the heat loads at cold temperature through all wind velocities is assumed to be the same as the rated condition. At low wind velocities, the efficiency is lower. The main loss contribution is the refrigeration loss at low wind speed. Increasing the wind velocities can increase the efficiency and the highest efficiency is 84.1%. Copper resistive loss, and refrigeration losses are the largest contributors.

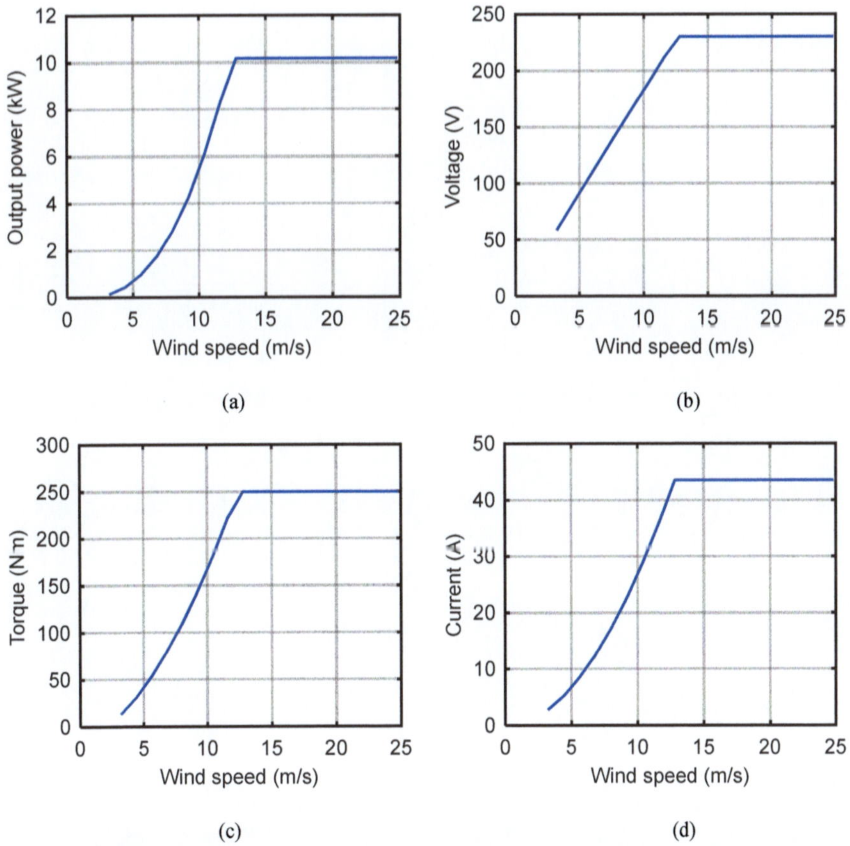

Fig. 5.25: Illustration of the power, voltage, current, and torque of the generator with different wind speeds, (a) Output power, (b) Voltage (c) Torque, and (d) Current.

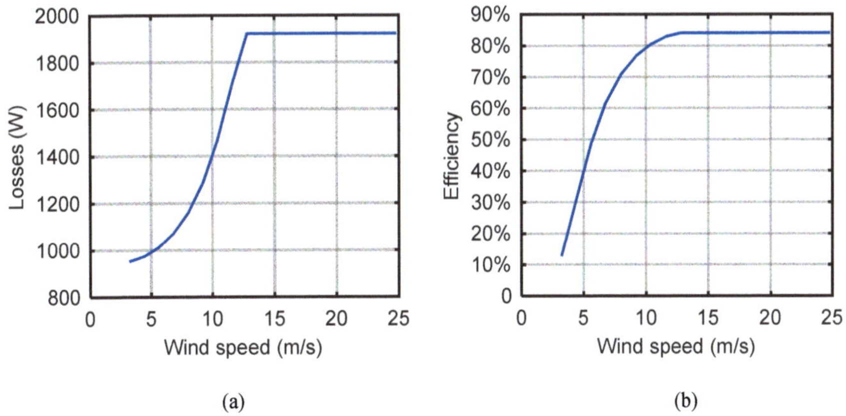

Fig. 5.26: Illustration of losses and efficiency of the generator with wind speeds, (a) losses, and (b) efficiency.

6 Preliminary tests for major components of the demonstrator

6.1 Critical current of the superconducting tapes

Superconductors exhibit no DC resistance within the regime bounded by critical temperature, critical current density and critical magnetic field. The critical current density is the upper limit of the current carrying ability within the superconducting state. The critical current density is calculated from the critical current divided by the area of the superconductor's cross section. The critical current is defined where the superconductor starts its phase transition. In practice, the critical current can be obtained when the voltage drop of the superconductor is 1 μV/cm, known as electric field criterion. Besides the electric field criterion, there are resistivity criterion and offset criterion to define the critical current in the data analysis. In this section, we choose the electric field criterion, which is the most used criterion [158]. To measure the critical current, a *V-I* curve, which is the voltage drop of the material as a function of applying current, is usually drawn and fitted to the power law expressed as in eq. (6.1) [67].

$$E(I) = E_c \left(I / I_c \right)^n \tag{6.1}$$

where $E(I)$ is the voltage drop across the superconductor, E_c is the electric field criterion, I is the current carried by the superconductor, I_c is the critical current, and n is the sharpness of the phase transition.

Superconducting coils are used in the generators mainly to generate the main flux density, which is one of the most critical and important parts of the power conversion. The operating temperature of the generators ranges from 10 K to 77 K. In a large-scale wind generator, the magnetic field in the superconducting coils can be up to 5.0 T, or even higher. In addition, orientation and amplitude of the flux density to each superconducting tape are different in a superconducting coil. At present, the typical manufacturer data for the HTS is not sufficient to make an accurate design, since the critical current of the superconductor with temperature, magnetic field and angular dependence is needed. To design the superconducting coil the characterization of the superconducting tapes is needed in the above mentioned temperature and field range. With that data a suitable superconducting coil for a wind generator can be designed.

Therefore, a measurement system is developed to characterize the critical current as a function of temperature, magnetic field and field angle. The system employs conduction cooling. To avoid Joule heat in the cryostat, pulsed current supply is used.

6.1.1 Measurement setup

The test system is shown in Fig. 6.1.

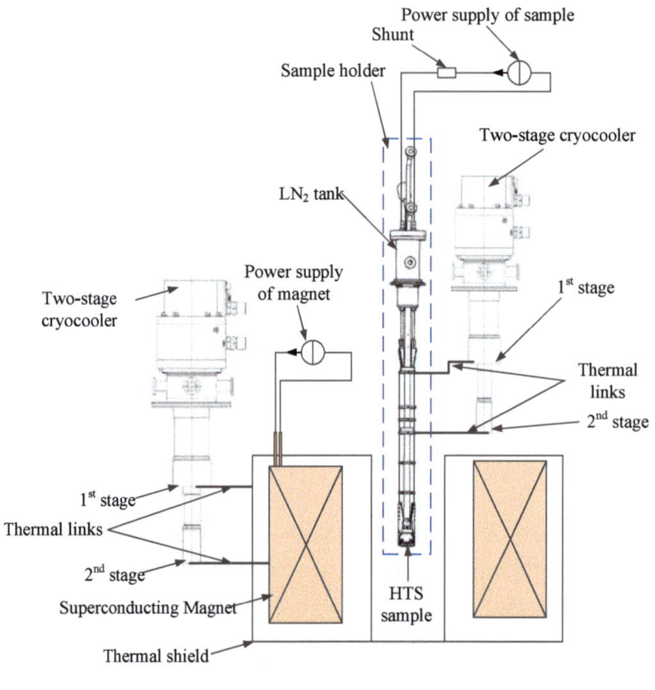

Fig. 6.1: Schematic overview of the system.

In order to measure the critical current of the HTS tapes under a high magnetic field, a superconducting magnet is used to produce the magnetic field from 0 T to 8 T and the operating temperature of the magnet is 4.2 K. The sample holder is not only to support the HTS tape, but also to set a specific angle to the magnetic field. In addition, the sample holder acts as a medium for cooling system and current input to the sample. Pulsed current supply up to 600 A is employed for the current input of the HTS sample. To cool down the HTS sample below 77 K, a two-stage cryocooler and helium gas are utilized, and liquid nitrogen is employed to cool down the copper current leads. The shunt is used to monitor the current of the sample power supply.

Cryostat CS210*F-GMX-19NP @ Advanced Research Systems is used for the sample holder, as shown in Fig. 6.2. A two-stage cryocooler is connected to the cryostat by two cooling links. The temperature of the 1st and 2nd stage cooling link is 77 K and 4.2 K, respectively. Besides the cryocooler, helium gas at a pressure of 50 mbar is employed in the cryostat to cool down the sample. Instead of a liquid coolant, helium gas acts as a cooling medium, which conveys the heat loads from the sample and current leads to the cryocooler. The cooling capacity of the cryocooler is summarized in Table 6.1. In order to further cool down the current leads, liquid nitrogen is employed in the sample holder to maintain the 77 K at the first stage of cooling.

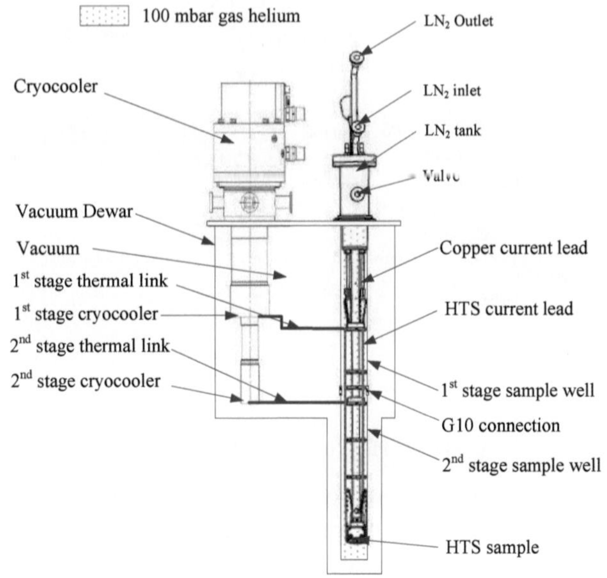

Fig. 6.2: Cryostat of the sample holder.

Table 6.1: Cooling capacity of the two-stage cyocooler of sample holder

Temperature	Cooling power	Unit
77 K	25	W
20 K	16	W
10 K	9	W
4.2 K	0.8	W

The photo of the sample holder is highlighted in Fig. 6.3. Cu-HTS current leads are employed in the sample holder to reduce the heat loads for the cryostat. The upper Cu-parts of the current leads are located in a liquid nitrogen tank to maintain a temperature of 77 K.

Outside the tank is vacuum insulation chamber surrounded by a stainless steel wall. In order to enhance the thermal link and contact surface between the holder and cryocooler, two Cu springs are employed. Besides, Cu-disks are employed to reduce the thermal resistance in between the helium gas and HTS current leads. G10 support and the Cu disks are also used to support the HTS current leads. A G10 tube is employed not only as support structure, but also as the channel for measurement wires. A moveable support for the HTS sample is necessary to study angular dependence of the HTS tape.

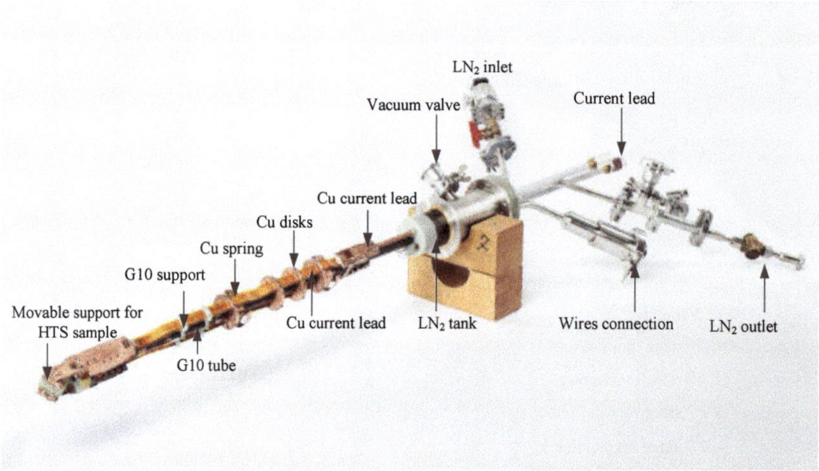

Fig. 6.3: Photo of the sample holder.

Three pairs of voltage taps are used to measure and monitor the voltage level of the sample and HTS current leads. To precisely measure the voltage across the sample, a Nano voltmeter Agilent 34420A is employed. It is also necessary to monitor the voltage drop of the current leads, as Cu-HTS current leads are used. Hence, Model 7700 switching module and Model 2700 multimeter to fulfil the task. To monitor and control the temperature of sample holder 8 temperature sensors are employed. Two PT 100s are used to monitor the liquid nitrogen level in the liquid nitrogen tank of the sample holder. Another two PT100s are located in the copper current leads. Four TVOs are installed to obtain the temperature information of the sample, first and second stage thermal link. To precisely control the temperature of the sample, Model 336 @ Lakeshore is employed and Model 218 @ Lakeshore is for temperature reading. Heat load is essential to control the temperature of the sample; hence, thermofoil heaters are used to provide the necessary heat load.

6.1.2 Test results

6.1.2.1 Test of sample holder

After the sample holder was inserted in the cryostat and 50 mbar gas helium was filled in the sample holder chamber, the sample holder was cooled down. The temperature variation during the cooling down process is drawn in Fig. 6.4. With helium gas, the sample is cooled down to 14.5 K after 6 hours and the current lead temperature is 276 K. In order to further cool down, liquid nitrogen is used to cool the current leads. After 2 hours, the temperature of the sample is stable at the setting temperature. The warm up process without heaters are illustrated in Fig. 6.5. It takes 2 days to warm the sample holder up without heater.

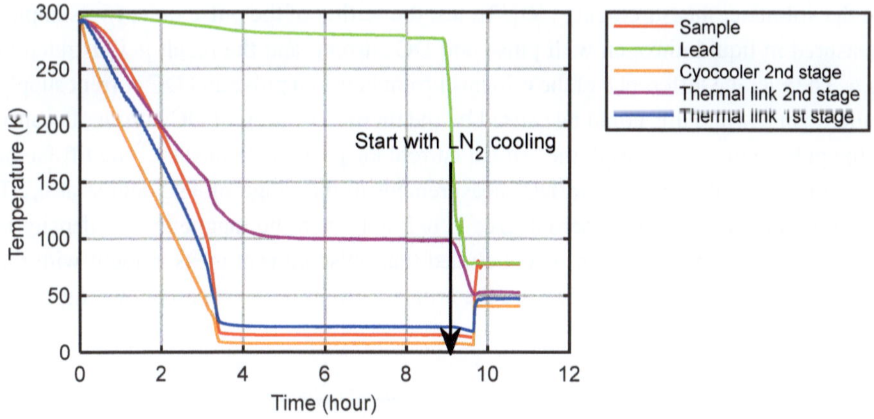

Fig. 6.4: Cooling down of the sample holder.

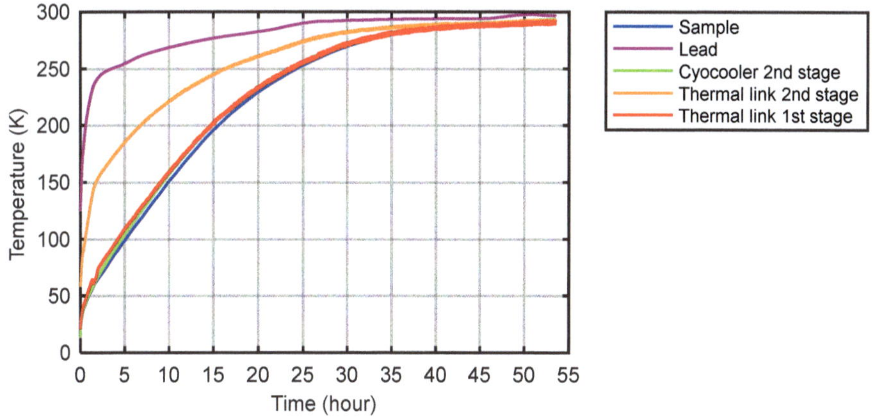

Fig. 6.5: Warm up of the sample holder.

6.1.2.2 Results of critical current measurement

When the current is high, the current-contact resistance is also high, which can affect the measurement result. To decrease the heating effect, pulsed measurement is employed. To correct for the voltage offset in the voltmeter and remove the non-negligible thermo-electric voltage effect, the sample voltage is measured at zero current before the pulse to dynamically cancel the offset voltage [158],[159]. During the pulse measurement, the current is increased from zero to the current level usually in 100 ms and then down to zero. In this process, there is an induced voltage when the current is rising and falling, which will last about a few hundreds of milliseconds [160].

The rising and falling time of the pulse current is designed to be 100 ms to have a relatively gentle slope. The constant current lasts 420 ms and it is assumed the voltage is stabilized before the voltmeter measurement [160]. To test the setting of the pulse time, two samples are measured in liquid nitrogen with pulse and DC current, and the result is illustrated in Fig. 6.6. The average difference of the critical current between pulse and DC current supply is smaller than 5% and this could be caused by the thermal heating of DC current, because when the pulse time is increased, the critical current keeps almost constant. The lift factor difference between the pulsed and DC measurement in the range of ±90 degree at 0.2 T and 0.5 T is shown in Fig. 6.7. Zero degree is defined when the magnetic flux density is parallel to the tape surface. It can be concluded that pulse current measurement with the test device is accurate enough to measure the critical current of the superconducting tapes.

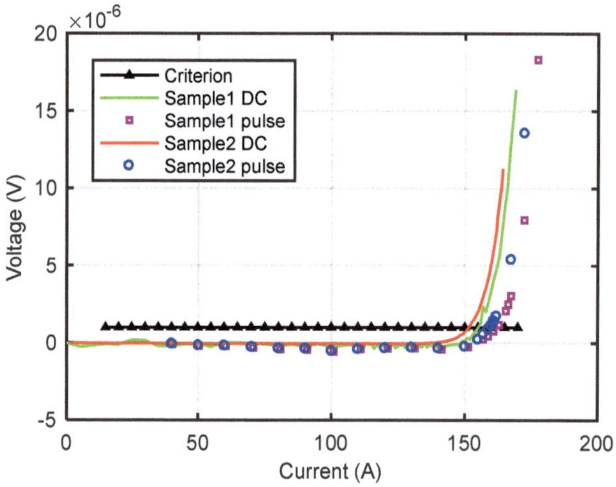

Fig. 6.6: Comparison of critical current in liquid nitrogen with DC current supply and pulse current supply.

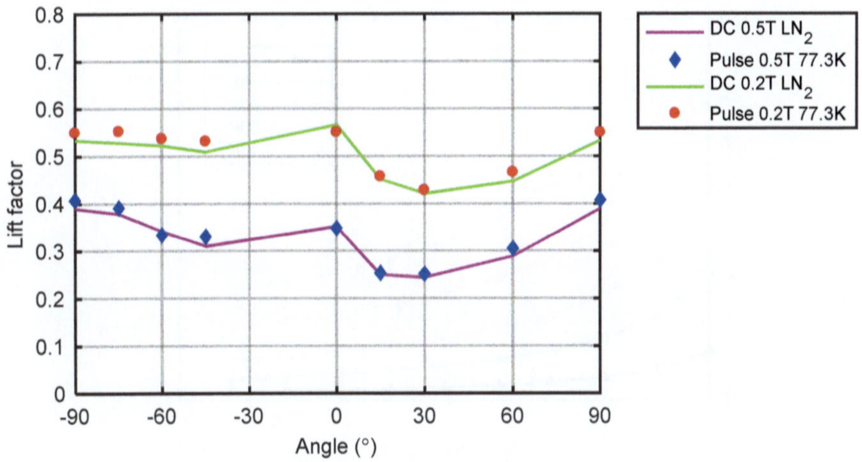

Fig. 6.7: Comparison of lift factor between pulse and DC current supply at 0.2 T and 0.5 T, respectively.

The critical current of the superconductor drops rapidly with increasing temperature and magnetic field. The temperature and magnetic field dependence of sample@ Superpower with the angle of 45° are illustrated in Fig. 6.8 and Fig. 6.9.

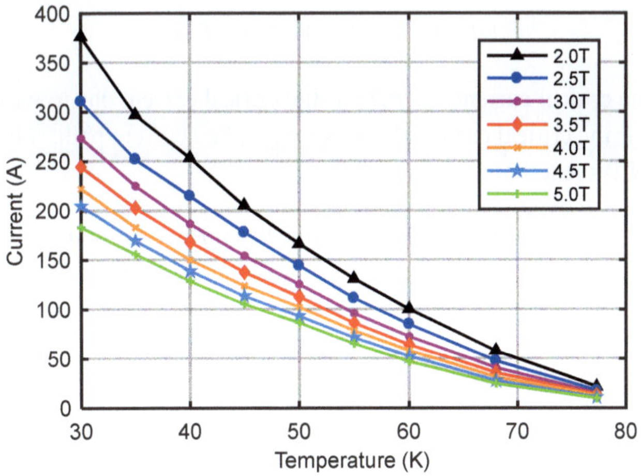

Fig. 6.8: Temperature dependence of the tape @ Superpower with the angle of 45°.

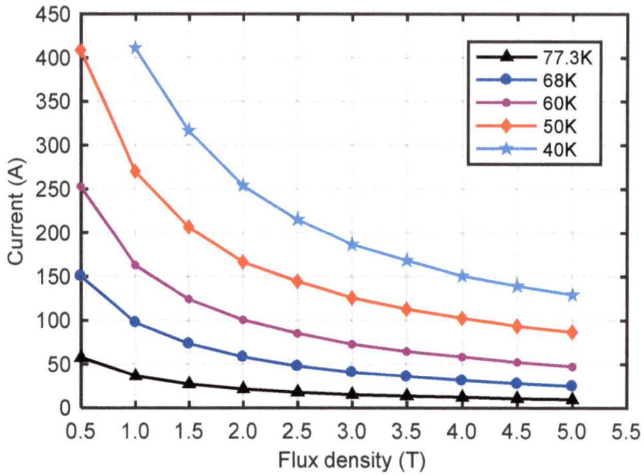

Fig. 6.9: Magnetic field dependence of the tape @ Superpower with the angle of 45°.

To describe the magnetic field dependence the model of two parallel current paths are used. One current path consists of grains connected by weak links which has a strong field dependence. Another current path is formed by strongly linked grains, which is determined by intragrain pinning. The total critical current is given by [161]:

$$I_c(B,T) = I_{cw}(B,T) + I_{cs}(B,T) \tag{6.2}$$

where $I_c(B,T)$ is the critical current, $I_{cw}(B,T)$ is the critical current due to weak linked current path, $I_{cs}(B,T)$ is the critical current due to strong linked current path, B is the magnetic flux density, and T is the temperature.

A phenomenological model similar to Power-Law dependence introduced by Kim is used to express the magnetic field dependence of the weakly linked critical current by [145]

$$I_{cw}(B,T) = \frac{I_{cw}(0,T)}{1+\left(\frac{|B|}{B_0(T)}\right)^{\beta}} \tag{6.3}$$

where, $B_0(T)$ depends on the inverse of the average grain size and the inverse of the London phenomenon, and β is related to the junction geometry. The strongly linked critical current is shown by eq. (6.4) based on collective pinning theory [162].

$$I_{cs}(B,T) = I_{cs}(0,T)\exp\left(-\frac{B}{B_{sc}(T)}\right)^{\alpha(T)} \tag{6.4}$$

where, $B_{sc}(T)$ is related to the irreversibility field and $\alpha(T)$ is used to take into account the thermally activated flux creep. $\alpha(T)$ is determined empirically from the slope of the critical current at high magnetic field with different temperatures.

A linear temperature dependence of $I_{cw}(0,T)$ and $B_0(T)$ in eq. (6.3) is observed [144],[163],[164]:

$$I_{cw}(0,T) = I_{cw}(0,0)\left(1-\frac{T}{T_c}\right) \tag{6.5}$$

$$B_0(T) = B_0(0)\left(1-\frac{T}{T_c}\right) \tag{6.6}$$

where, T_c is the critical temperature of the superconductor.

The temperature dependence of $B_{sc}(T)$ and $I_{cs}(0,T)$ in eq. (6.4) are also given in [144]:

$$B_{sc}(T) = B_{sc}(0)\exp\left(-\frac{T}{T_{sc}}\right) \tag{6.7}$$

$$I_{cs}(0,T) = I_{cs}(0,0)\left(1-\frac{T}{T_c}\right)^{\gamma} \tag{6.8}$$

where, T_{sc} and γ are parameters and T_{sc} has a unit of K.

Depending on experimental results, Reference [165] proposed a very simple scaling law for $I_c(B,T)$ below 40 K, as shown in equation (6.9). Different from eq. (6.2), eq. (6.9) has no physical meaning and extracted from empirical results. This equation is simple and less parameters are needed, but has larger calculated error.

$$I_c(B,T) = I_c(0,0)B^{-\alpha}\exp\left(-\frac{T}{T^*}\right) \tag{6.9}$$

Where α and T^* are parameters, which are determined by empirical data.

Besides magnetic field and temperature dependence, the critical current of the superconductor is influenced by the external field direction, which is called angular dependence.

Generally, the critical current has a much higher critical current when the magnetic flux density is aligned with the crystallographic ab-planes of the YBCO layer and a lower value when the magnetic flux density perpendicular to the wider surfaces of the tape. However, the angular dependence of the superconducting tapes depending on manufactures, and the higher critical current is reached at different angles. In Fig. 6.10 the angular dependence of the tape with advanced pinning from superpower at 77.3 K and 68 K is illustrated. Zero angle is defined as the flux is parallel to the wider surface of the tape.

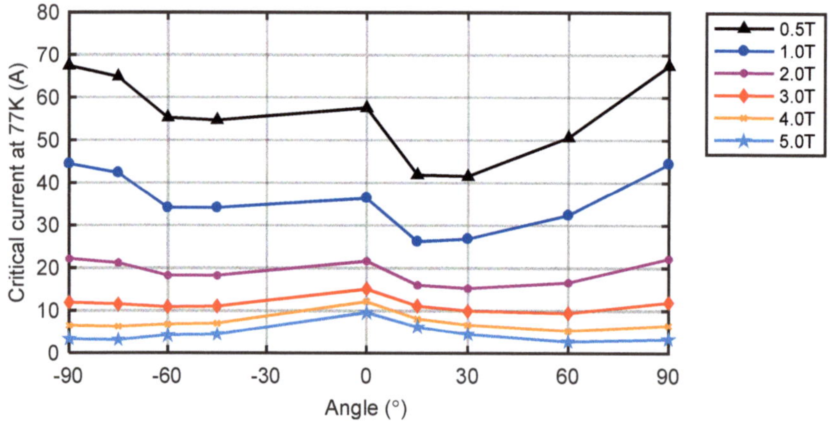

(a)

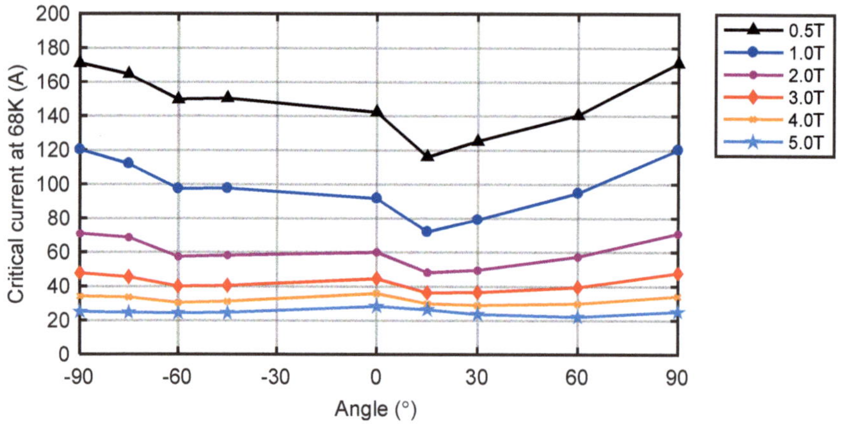

(b)

Fig. 6.10: Angular dependence of the tape from Superpower at 77.3K and 68K, respectively.

From Fig. 6.10 it can be seen that the lowest critical current of the sample is not with perpendicular field to the wider surface. When the magnetic field and temperature is higher, the angular dependence becomes less prominent. Moreover, the behaviour of the critical current with perpendicular field also changes with magnetic field: at lower field the critical current is higher with perpendicular field than with parallel field, which is contrary to the behaviour at high magnetic field.

The critical current in an anisotropic superconductors with isotropic pinning sites is well described by elliptical angular dependence, as expressed by eq. (6.10) [166]-[168],

$$f(\theta)=\sqrt{\cos^2\theta+u^2\sin^2\theta} \tag{6.10}$$

where the origin of the angle θ is for an applied field parallel to the ab planes.

Integrated the elliptical angular dependence into the generalized Kim model without taking the temperature into consideration, the angular dependence influence on the critical current can be derived as

$$I_c(B)=\frac{I_0}{1+\left(\frac{f(\theta)B}{B_0}\right)^{\beta}} \tag{6.11}$$

where I_0, B_0, β are parameters.

To describe the regular dependence on the orientation of the magnetic field, with the maximal critical current when the magnetic field is parallel to the tape and minimal critical current when the magnetic field is perpendicular to the tape, eq. (6.10) and (6.11) is used. However, for more complicated angular dependence, for instance, the angular dependence of Superpower tape with advanced pinning, eq. (6.10) and (6.11) is not enough. In that case, three overlapping elliptical dependences based on the observed peaks in the angular dependences of critical current are proposed [146],[166] with up to 15 parameters. Even though it is true that all the parameters can be extracted from experimental results at low field (up to 400 mT) [146],[166],[169], it is impossible to use the same parameters to describe angular dependence of the critical current for the whole field range of 0-5 T. For engineering application, magnetic field can be classified to different range cases and within each case the same angular dependence parameters are employed.

6.2 Test of the iron materials

The silicon lamination sheets help to reduce the reluctance in the magnetic circuit and confine the flux in the generator. The amount of superconducting tapes can be decreased significantly by employing iron rotor core and stator core. Moreover, the force acting on the superconducting coil is reduced if an iron pole is used. To simplify the cryogenic cooling system and supports structure of the superconducting coils, iron cores can be designed at the same working temperature as superconducting coils. There are two iron core topologies for superconducting coils of the superconducting wind generators, cold iron and warm iron topology [91]. The cold iron design offers a more stable cryogenic temperature as the cold iron contributes to the cold mass and enhances the thermal capability of the system. However, the iron has higher losses to harmonics. Hence, cold iron needs higher cooling power and longer cool down time [96]. Whether to have cold iron or warm iron concept for a certain design requires a trade-off study on superconducting cost and losses at cryogenic temperature. To start this study, the data of the ferromagnetic materials are necessary not only at room temperature but also at cryogenic temperature. However, few cryogenic data of the magnetic materials are available for designers to use. In order to fill this gap, cooperating with Institute of Electrical Engineering (ETI), Hybrid Electric Vehicles (HEV), we carried out the measurement to study the magnetic properties at room temperature and liquid nitrogen temperature (77 K).

6.2.1 Measurement system

Toroidal cores are used to identify the magnetic properties based on the international standard IEC 60404. A ring specimen is chosen to simplify the structure at 77 K. The ring specimen of the lamination sheets is 45 mm in inner diameter and 55 mm in outer dia-meter. The structure is shown in Fig. 6.11. The samples tested are M330-35A, M330-50A, M235-35A, and VACODUR 49 (cobalt iron), which covers different grades, thicknesses and saturation points. The detailed information of the tested samples is summarized in Table 6.2. Please note that 2.35 T is the saturation point of Vacodur49.

The system overview is shown in Fig. 6.12. The excitation of the primary winding of the ring specimen is powered by a high performance linear amplifier that consists of multiple parallel amplifier modules. A field programmable gate array (FPGA) based high speed control algorithm is used to maintain equal output current for each module. The algorithm calculates individual input voltages for the amplifier modules, which are supplied by a digital-analog-converter. In order to precisely measure the power loss of the test specimen, the secondary induced voltage $u_{s,ind}$ has to be controlled to a sinusoidal waveform with a total harmonic content as low as possible. This controller is also implemented on FPGA to realize fast sample rates and control times with quality through an analog-digital-converter.

Detailed system introduction is shown in [170]. Based on Ohm's law and Faraday's induction law, the power is absorbed by the toroidal core is a composition of the ohmic losses in the primary windings and the core losses in the toroidal core, as expressed in eq. (6.12).

Table 6.2: Specification of magnetic materials tested at cold temperature

	M330-35A	M330-50A	M235-35A	Vacodur49
Density	7620kg/m^3	7650kg/m^3	7600kg/m^3	8120 kg/m^3
Thickness	0.35 mm	0.50 mm	0.35mm	0.27 mm
Conductivity at room temperature	2.29 × 10^6 S/m	2.33× 10^6 S/m	1.92× 10^6 S/m	2.35× 10^6 S/m
Conductivity at 77 K	2.90× 10^6 S/m	2.93× 10^6 S/m	2.11× 10^6 S/m	2.59× 10^6 S/m
Min.Polarisation (10,000 A/m)	1.7 T	1.7 T	1.7 T	2.35 T

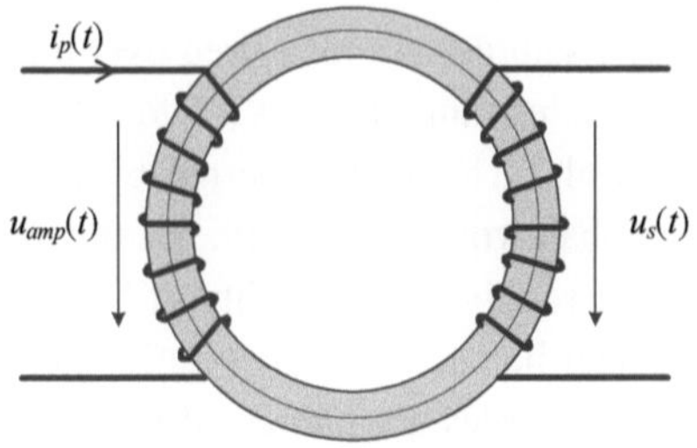

Fig. 6.11: Toroidal cores of the measurement.

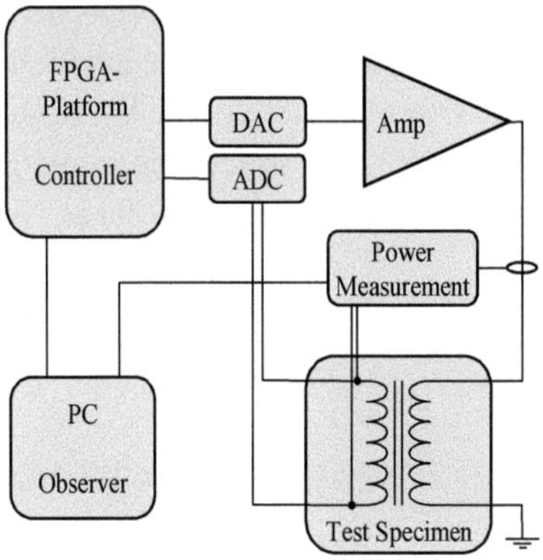

Fig. 6.12: Sketch of measurement system.

$$\underbrace{u_{\mathrm{amp}}(t)i_{\mathrm{p}}(t)}_{A}=\underbrace{R_{\mathrm{p}}i_{\mathrm{p}}^{2}(t)}_{B}+\underbrace{\frac{N_{\mathrm{s}}}{N_{\mathrm{p}}}u_{\mathrm{s,ind}}(t)i_{\mathrm{p}}(t)}_{C} \tag{6.12}$$

where, $u_{\mathrm{amp}}(t)$ is the amplifier voltage, $i_{\mathrm{p}}(t)$ is the primary winding current, R_{p} is the resistance of the primary winding, $u_{\mathrm{s,ind}}(t)$ is the induced voltage of the primary winding, N_{s} is number of turns of secondary winding, N_{p} is the number of turns of primary winding, term A is the input power to the toroidal core, term B is the ohmic losses in the primary winding, and term C is core magnetization and losses. From term C it can be seen that the core magnetization and losses can be determined by observing the secondary voltage and the primary current over time. The reactive power of term C is the magnetization of the core and the active power of it is the core losses.

6.2.2 Measurement results

The magnetic properties of the samples are measured over a frequency range from 50 Hz to 1000 Hz and and a flux densities range from 0.1 T to 2.1 T for the Vacodur49, 0.1T to 1.5 T for the other three materials.. The relative permeability curves of all the samples are illustrated in Fig. 6.13. The cobalt iron, Vacodur49, has the highest relative permeability, followed by the M330, and the relative permeability of M235 is lowest at room temperature (RT). From Fig. 6.13 it can be seen that the relative permeability is increased with higher flux densities when the samples are immersed in liquid nitrogen. From the design point of view of the superconducting generators, this is an advantage, as less volume of superconductors is used to produce the same air gap flux density. The increase of the relative permeability of the different samples at 77 K compared with room temperature at 50 Hz is also shown in Fig. 6.13. The M235 material has the largest temperature dependence on relative permeability and cobalt iron has the smallest. The maximum increase of the relative permeability of M235 is 63%, 48% of the M330 and 10% of Vacodur49. Moreover, the maximum increase seems to be around the saturation point of the materials at room temperature, which leads to higher saturation point at 77 K than at room temperature.

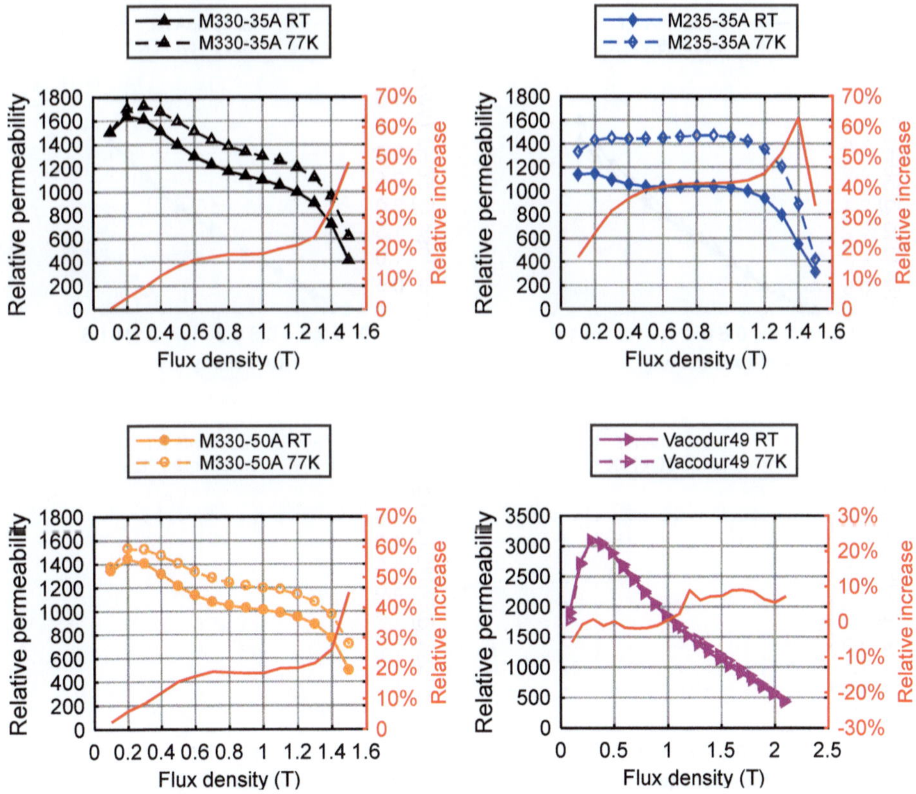

Fig. 6.13: Relative permeability of the four different samples at 50 Hz.

Besides the permeability, iron losses of the samples are also investigated. The iron loss of the four samples at 50 Hz is shown in Fig. 6.14. Different materials have a different loss behavior when cooled down to 77 K. In general, the lower the temperature, the higher the losses. The iron loss of the materials consists of hysteresis loss, classical eddy current loss and excess loss, according to Bertotti [171],[172]. Below 400 Hz the dominant part of the iron losses are expected to be hysteresis loss and excess loss for M235-35A, M330-35A, and Vacodur49. Hysteresis loss and excess loss are related to the statistical distribution of the local coercive fields and the grain size, as well as flux density and frequency [173]. As for the M330-50A, the dominant losses should be hysteresis loss and eddy current loss, which is caused by a larger thickness. The increase of iron losses of the four materials at 50 Hz over flux density range are also summarized in Fig. 6.14. The M235 sample has negative temperature dependence on iron loss when the flux density is lower and positive temperature dependence with higher flux density, whereas the other samples all has positive temperature dependence on iron loss.

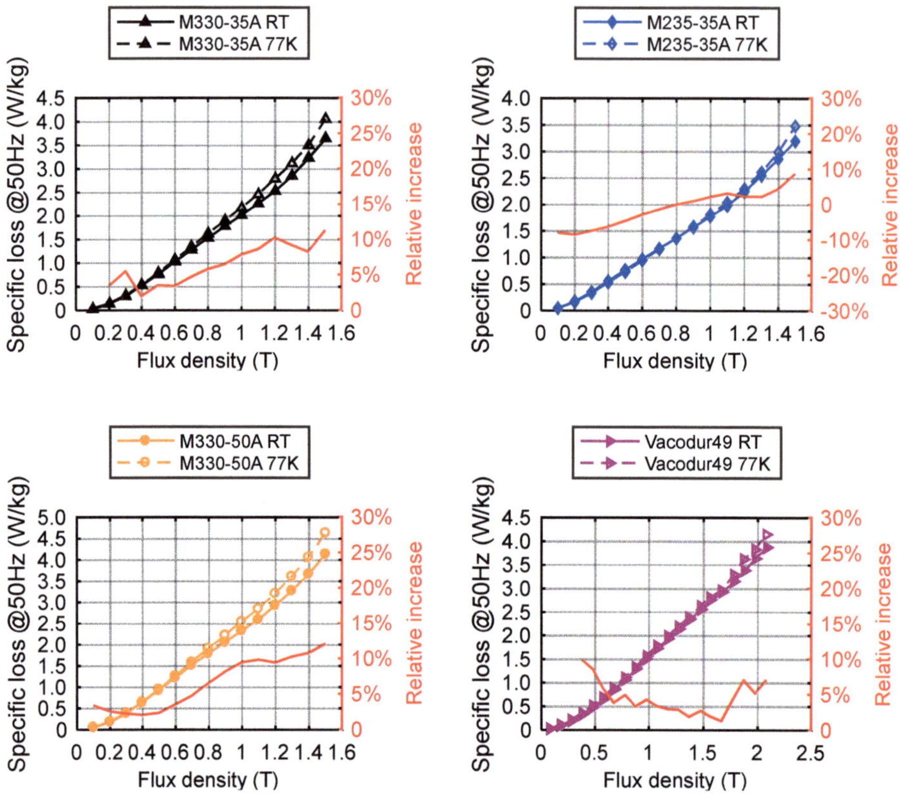

Fig. 6.14: Iron loss of the four different samples at 50 Hz.

Iron losses over a range of frequency are compared with the four materials at flux density at 1.5 T, as illustrated in Fig. 6.15. At higher frequency, the dominant losses are excess loss and eddy current loss. The relative increase of iron losses of the four materials at 1.5 T over frequency are shown in Fig. 6.15 too. Over the frequency range, the temperature dependency is positive. After 200 Hz, the loss increase from room temperature to 77 K is relatively steady. The average increase of iron loss when frequency is higher than 200 Hz at 1.5 T is 11% for M235-35A, 18% for M330-35A, 20% for M330-50A, and 9% for Vacodur49.

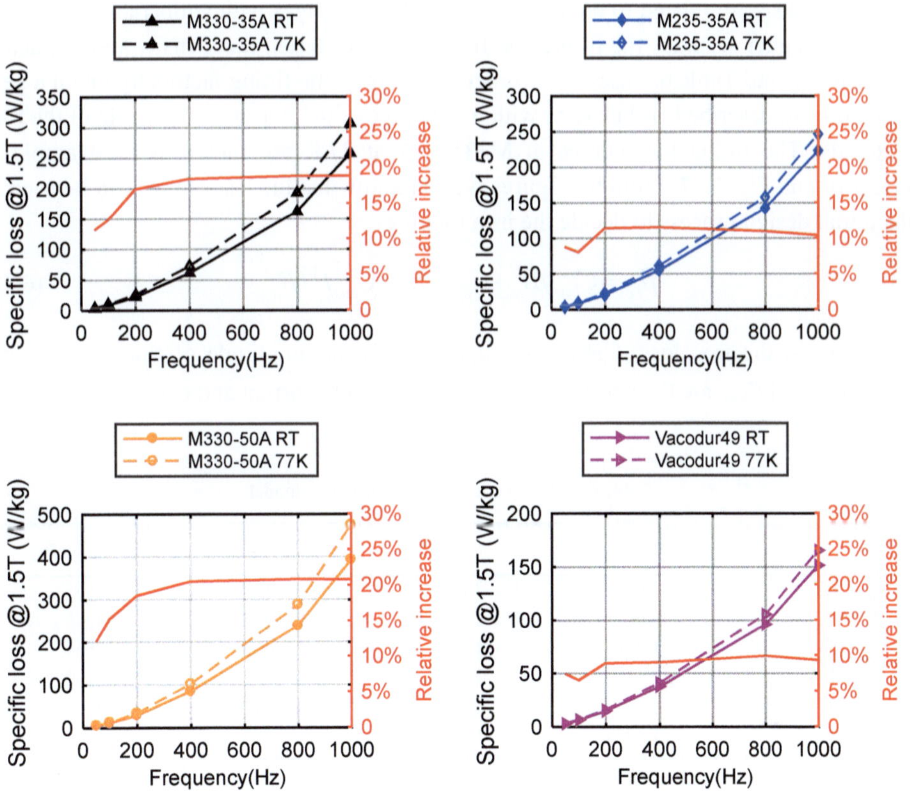

Fig. 6.15: Iron loss of the four different samples at 1.0 T.

6.2.3 Iron loss model

Statistically, the fundamental frequency of the direct-drive superconducting wind generators is low, usually less than 10 Hz. Generally, in a three phased generator with air gap armature winding, the main spatial magnetomotive force harmonics are of order 5 and 7, 11 and 13. Hence, we can investigate the fitting curve of the iron loss between 50 Hz and 400 Hz to carry out the initial design of the rotor iron topology and decide whether to locate the iron at cryogenic or room temperature.

When the skin effect is negligible, usually less than 400 Hz [174], Bertotti developed an iron loss model as expressed in eq. (6.13) to describe the loss behaviour with flux density, frequency and empirical factors. These fitting factors to describe the frequency range of 50 Hz to 400 Hz are listed in Table 6.3. From Table 6.3 it can be seen that the temperature mainly has an influence on the eddy current and excess coefficients, and hysteresis coefficients remain almost the same. The reason is that the coefficients of the eddy current loss and excess loss are related to the conductivity which changes with temperature. To reach relatively higher accuracy the fitting coefficients with different flux density and frequency are different and Table 6.3 uses the average value of all the fitting factors to obtain a general model to describe loss behavior within the range of 0-1.5 T and 50-400 Hz of the four materials. The fitting loss curves of M235-35A at 77 K are shown as an example in Fig. 6.16 and Fig. 6.17. From the figures we can know that the loss curves can be used in the initial design process to decide the iron topology.

$$p_{Fe} = C_{hyst} f B_p^2 + C_{ec} f^2 B_p^2 + C_{exc} f^{1.5} B_p^{1.5} \tag{6.13}$$

where p_{Fe} is the iron loss, f is the frequency, B_p is the amplitude of flux density, and C_{hyst}, C_{ec},and C_{exc} are the coefficients of hysteresis,eddy current and excess loss.

Table 6.3: Fitting coefficients of iron loss model

	M330-35A	M330-50A	M235-35A	Vacodur49
C_{hyst} RT	2.62E-2	2.89E-2	2.00E-2	2.03E-2
C_{ec} RT	6.06E-5	1.25E-4	5.09E-5	2.84E-5
C_{exc} RT	1.24E-3	1.18E-3	1.46E-3	9.08E-4
C_{hyst} 77K	2.64E-2	2.89E-2	2.09E-2	2.07E-2
C_{ec} 77K	7.67E-5	1.58E-4	5.59E-5	3.12E-5
C_{exc} 77K	1.62E-3	1.56E-3	1.73E-3	1.04E-3

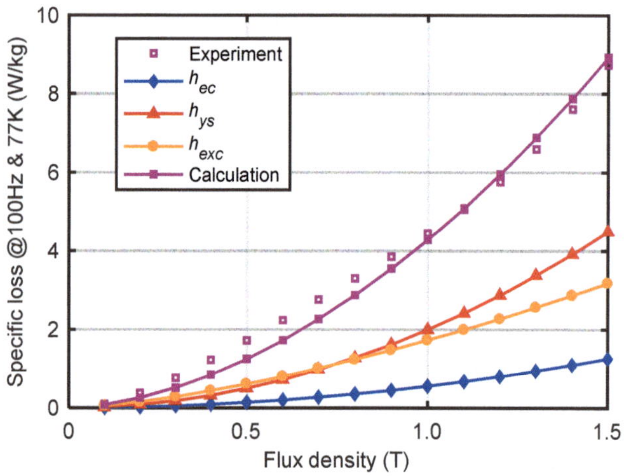

Fig. 6.16: Iron loss model of M235-35A at 100 Hz and 77 K (h_{ec} is the eddy current loss, h_{ys} is the hysteresis loss, and h_{exc} is the excess loss).

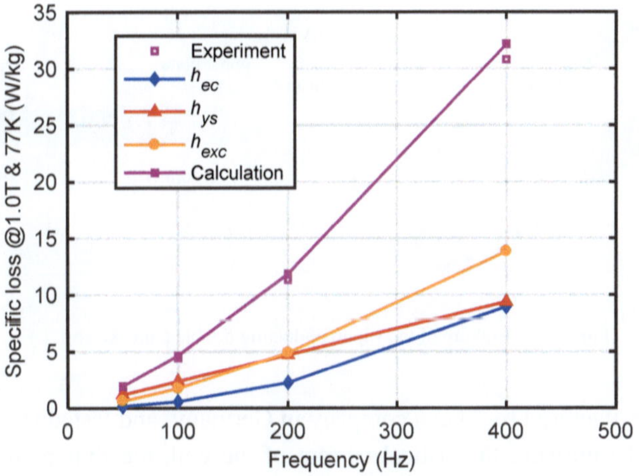

Fig. 6.17: Iron loss model compared with experiment data of M235-35A at 1.0 T and at liquid nitrogen temperature (h_{ec} is the eddy current loss, h_{ys} is the hysteresis loss, and h_{exc} is the excess loss).

6.3 Test of the superconducting coil

The superconducting coils in the demonstrator are located in the stator and generate the main flux density in the air gap, which is significantly important for the energy conversion. The performance of the superconducting coils plays an essential role in determining the generator performance. Hence, in-depth investigations on the superconducting coils are necessary before manufacturing the demonstrator. For a preliminary study, we tested the coil in liquid nitrogen instead of 30 K to investigate its performance.

6.3.1 Measurement result

Based on the design of the demonstrator, a double pancake superconducting coil was built. The structure of the coil is shown in Fig. 6.18 and the specification of the coil is summarized in Table 6.4. The double pancake coil consists of 244 turns of GdBaCuO tapes. The top and bottom coil are insulated by a 3 mm-thick Nomex-Kapton-Nomex. The turn to turn insulation is cowound with the YBCO tapes on a coilformer, which is used for mechanical stabilization. The layer to layer connection is achieved by soldering a piece of 12 mm GdBaCuO tape. Afterwards, the coil is wrapped by the resin. Two copper plates are placed at the top and bottom to enhance the thermal conductivity.

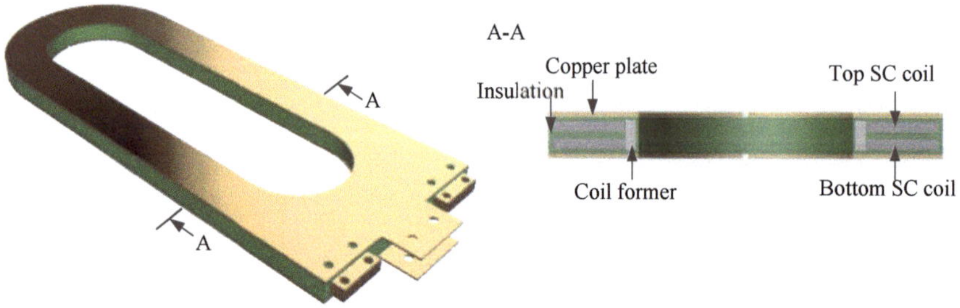

Fig. 6.18: Structure of the superconducting double pancake coil.

The coil was manufactured by Theva company in Germany, and tested at the company in liquid nitrogen. To measure the critical current of the coil, the four point measurement technique is employed. The current is slowly ramped up at the terminals of the coil, and the voltage between the terminals are recorded. When the voltage reached the given value of the electrical field(1 μVcm^{-1}), the corresponding current is the so-called critical current. Hence, to measure the voltage of the double pancake coil, we soldered three voltage taps: two in the terminals and one in the middle of the whole coil. The tested *V-I* curve of the superconducting double pancake coil in liquid nitrogen is illustrated in Fig. 6.19. The critical current is 48.6 A, by neglecting the offset voltage of the end to end connection. It can be seen from Fig. 6.19 that when the top coil reaches the critical current, the bottom coil is still under superconducting state, that is, the top coil decides the critical current.

Table 6.4: Specification of the superconducting double pancake coil

Parameter	Value
Tape	GdBaCuO
Tape thickness	0.22 mm
Tape width	4 mm
Copper stabilizer	100 um
Number of turns	2×122
Thickness of copper plate	2 mm
Height of the coil	19 mm
Inner/outer radius of the coil	45/83 mm
Turn to turn insulation	16 um PET Polyester insulation foil
Layer to layer insulation	Nomex-Kapton-Nomex
Layer to layer connection	100 mm long 12 mm width tape
Number of tape pieces	10
Joint resistance	< 30nOhm per joint
Total length of tapes	218 m

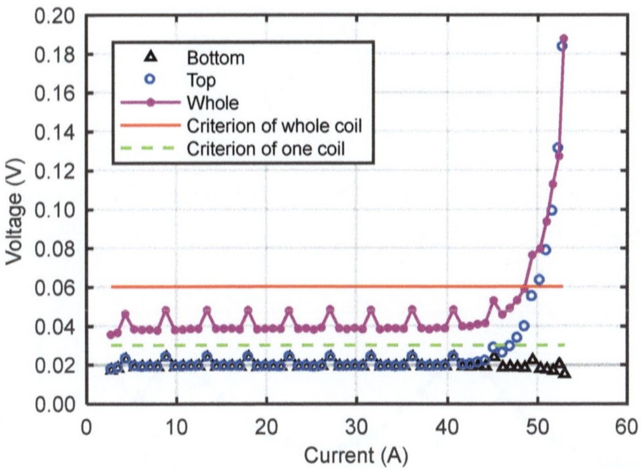

Fig. 6.19: Tested *V-I* curve of the superconducting double pancake coil in liquid nitrogen.

6.3.2 Simulation and discussion

To understand the measurement result, simulation models are employed. By integrating the critical current of the superconducting tape and the flux density into the *E-J* power law, the electric field E_j of the *j*th turn is calculated as eq. (6.14) [175]. Hence, the averaged electric field E_{avg} measured from end to end of the coil is as expressed in eq. (6.15). When the averaged electric field E_{avg} reached the critical electric field E_c, the current is defined as the critical current. To calculate this critical current, 2D analysis in Comsol is used. To obtain the critical current, it is necessary to know the the angular and field dependence of the superconductor. Hence, the angular and field dependence of the superconducting tape at 77 K is measured as shown in Fig. 6.20. From Fig. 6.20 it can be seen that the highest critical current is obtained at the angle of 150°and 330° instead of parallel field (0°, 180°, 360°). The lift factor is defined as the ratio of the critical current at a certain orientation and amplitude as well as temperature to the critical current at 77 K with self field. The minimal critical current of the tape at 77 K self field is 100 A.

$$E_j = E_c \frac{J}{J_c(\boldsymbol{B})} \left| \frac{J}{J_c(\boldsymbol{B})} \right|^{n-1} \tag{6.14}$$

where, E_j is the electric field of *j*th turn, $\boldsymbol{B}$ is the magnetic field vector to the superconductor, J is the suppled current density due to operating current, E_c is the critical electric field, n indicates the transaction sharpness from superconducting state to the normal state.

$$E_{avg} = \sum_{j=1}^{m} E_j l_j \Big/ \sum_{j=1}^{m} l_j \tag{6.15}$$

where, E_{avg} is the averaged electric field of the coil, m is the number of turns in one coil, l_j is the length of the jth turn.

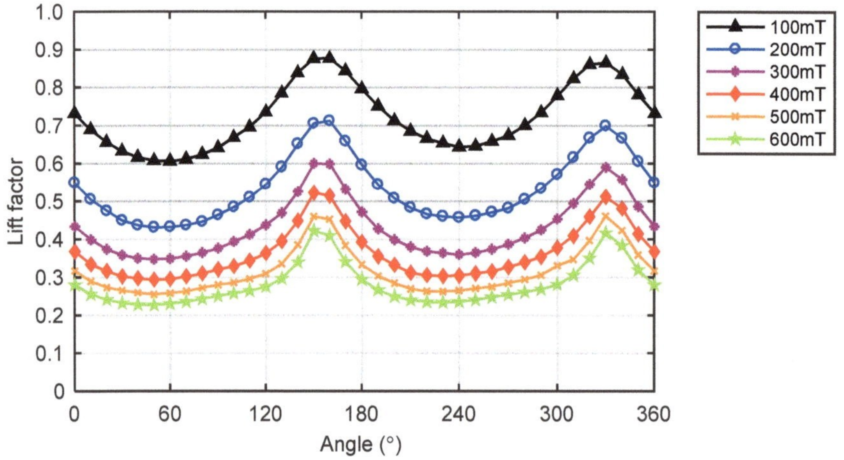

Fig. 6.20: Measured lift factor of the superconducting tape used in the coil in liquid nitrogen.

Through simulation, the critical current is 45.9 A, as shown in Fig. 6.21, which is consistent with the experiment result (5% difference). Generally, for a double pancake coil, the two pancake coils are symmetrical in dimensions, magnetic flux density and current density distribution. Moreover, the innermost turns of the two pancake coils experience the peak magnetic flux density and this region will develop a critical electrical field (1 μVcm^{-1}) first. However, when the exact angular dependence of the tape is considered, this critical region may differ.The current density distribution of the coil is shown in Fig. 6.22. Through Fig. 6.22 it can be seen that the two coils have an unsymmetrical current density over critical current density distribution due to the field and angular dependence of the tape. In addition, the innermost turns of the top coil reaches the critical current first, consistent with the measurement result in Fig. 6.19, which indicates the top coil reaches the criterion earlier than the bottom coil.

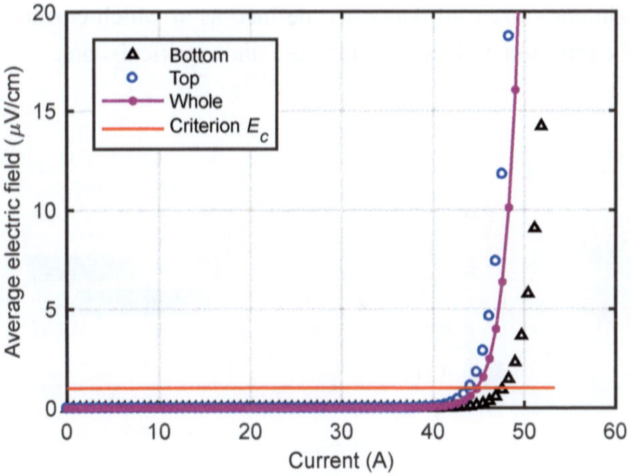

Fig. 6.21: Simulated average electric field as a function of the supply current.

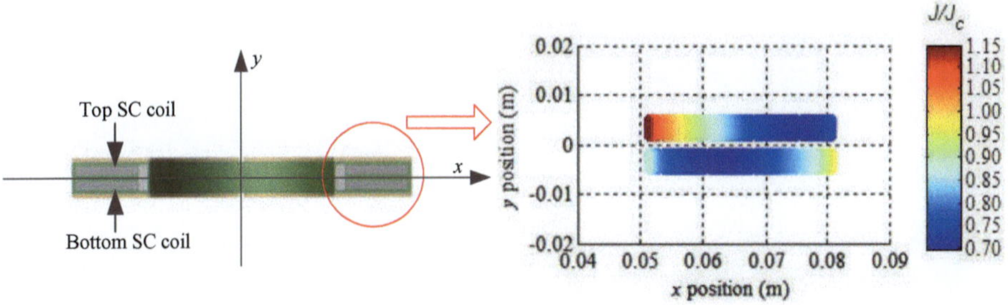

Fig. 6.22: Current density distribution of one superconducting coil side under critical current of 45.9 A.

Due to the average effect of the electric field of all turns in the superconducting coils, as illustrated in eq. (6.15), when the voltage between the terminals are used to measure the critical current, the innermost turn of the top coil has already run at over-critical state, as highlighted in Fig. 6.22.This is dangerous to the coil to run at even slight lower than the critical current of 45.9 A. Hence, to decide the maximum current the coil can carry without risk, it is essential to define this limit by the voltage of the innermost turn of the top coil.Consequently, the critical current is reduced to be 42.3 A. The flux density and current density distribution is presented as in Fig. 6.23. It can be seen that the maximum flux densityis 0.2 T and the flux density distribution is symmetrical. However, the consequenting current density over critical current density distribution is unsymmetrical due to the tape characteristics. Moreover, with current of 42.3 A, each turn of the whole coil is under critical-state. Therefore, it is highly recommended to operate this coil under the current defined by the innermost turn of the top coil to avoid any quench or risk. To conclude it more generally, the maximum current of the coil is defined as at which current there is a least one turn reaches the criterion and the rest turns are under critical-state.

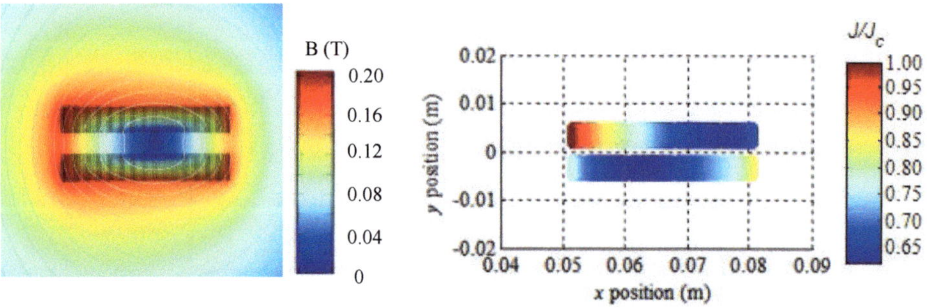

Fig. 6.23: Flux and current density distribution of one superconducting coil side under critical current of 42.3 A.

7 Design of 10 MW generators

To study the feasibility of a superconducting DC generator, a 10 MW generator is designed and compared to a permanent magnet and copper coil excited synchronous generator with the same rating. The main items to be investigated are torque to volume ratio, torque to weight ratio, efficiency, technology availability, reliability, installation and transportation cost and lifetime cost. This chapter will cover the first three and then the influence of the main design parameters, such as pole pair number, air gap diameter, and excitation material consumption, are discussed.

7.1 Design of a superconducting DC generator

After integrating the electromagnetic design model with mechanical design model, the superconducting DC generator is designed and optimized in an analytical way. By using normalization and weighting factors, the optimization objective function is built as shown in equation (2.34). This design and optimization method can calculate and evaluate hundreds of designs and find out the optimized design within hours, which is much faster than FE analysis software.

The DC generator employs radial flux structure and the rated power is 10 MW at 10 (1/min). The rated voltage of the DC generator is 1.0 kV. The air gap of the generator is selected to be 10 mm, taking into consideration the mechanical constraint and manufacture tolerance. The ranges of the six control parameters are listed in Table 7.1. To make each design of the generator reasonable, more constraints are needed. The maximum and minimum length of the superconducting tapes, mass, volume, and efficiency are limited to 1200 km, 26 km, 300 ton, 60 ton, 120 m^3, 20 m^3, 94% and 99.

Table 7.1: Design control parameters of a 10 MW superconducting DC generator

Symbol	Explanation	Range
D_a	Outer diameter of rotor	4-10 m
h_{sc}	Height of the superconducting coil	4-200 mm
h_{ry}	Height of rotor core	50-600 mm
h_{sy}	Height of stator core	50-600 mm
A_a	Electrical loading of armature winding	20-100 kA/m
p	Pole pairs number	10-40

The optimization process runs in terms of the objective function, minimal superconducting wire length, minimal mass, minimal volume and maximum efficiency. The parameters of the five designs are underlined in Table 7.2, where w_{sc} is the width of one superconducting coil side, and the design results are illustrated in Fig. 7.1. Each optimal process analyzed 1500 designs before the optimized one is chosen. The average time to locate one optimal design is 43 minutes.

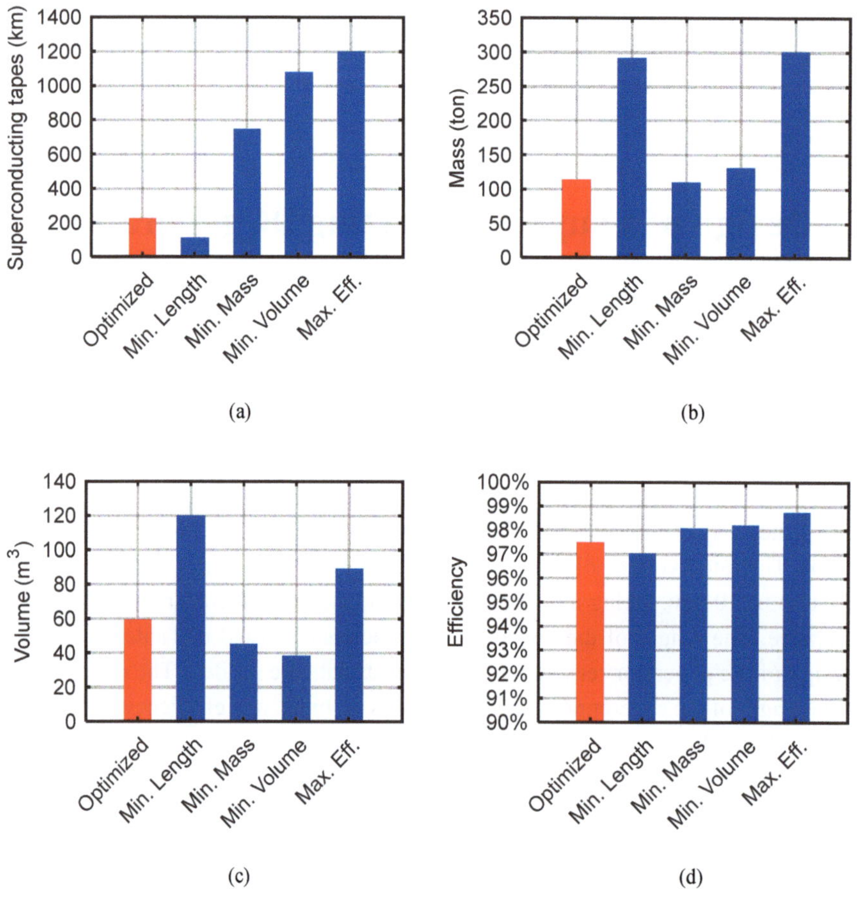

Fig. 7.1 Optimization results based on different objectives, (a) minimal superconducting wire length, (b) minimal mass, (c) minimal volume, and (d) maximum efficiency.

After the optimal design is located by analytical methods based on the objective function, the detailed design by FE analysis is made to validate and investigate its performance. According to the design results above, a two-dimension FE model is built. The no-load flux density distribution of one pole pair model is illustrated in Fig. 7.2. The flux density of the armature winding in the middle radial position is shown in Fig. 7.3, as well as the harmonic spectrum. The magnitude of the fundamental flux density is 2.3 T and the average flux density is 1.5 T. In this model, YBCO is used to make the superconducting field coils. The critical current of the superconducting tapes is 130 A @ 77 K/self-field. The working temperature of the superconducting coils is 30 K. The maximum perpendicular field in the superconducting coil is 2.63 T, which decides the operating current of the coil. A lift factor of 2.54 at this magnetic field can be assumed from [176]. With a safety factor of 0.65, the operating current is selected to be 214 A. The design parameters and specifications of the generator are listed in Table 7.3 .

Table 7.2: Optimal designs of the superconducting DC generator

Parameter	Objective function	Min. superconductor length	Min. mass	Min. Volume	Max. eff.
D_a	5.49 m	8.79 m	5.40 m	4.03 m	8.80 m
h_{sc}	9 mm	4 mm	47 mm	63 mm	102 mm
w_{sc}	101 mm	77 mm	59 mm	76 mm	39 mm
h_{ry}	195 mm	545 mm	244 mm	388 mm	600 mm
h_{sy}	194 mm	181 mm	278 mm	414 mm	409 mm
A_a	97.3kA/m	91.2kA/m	99.0kA/m	99.5kA/m	45.6kA/m
i_{sc}	214 A	282 A	125 A	105 A	131 A
p	14	24	21	14	36

Table 7.3: Design specifications and parameters of the 10 MW superconducting DC generator

Parameter	Value	Unit
Rated power	10	MW
Rated speed	10	1/min
Rated voltage	1	kV
Number of poles	28	-
Number of coils	1176	-
Stator outer diameter	6034	mm
Stator inner diameter	5513	mm
Stator core thickness	194	mm
Physical air gap	10	mm
Rotor outer diameter	5493	mm
Rotor inner diameter	4908	mm
Rotor core thickness	195	mm
Stack length	1450	mm
Electrical loading	97.3	A/mm
Average flux density	1.5	T
Number of turns per superconducting coil	1857	-
Operating current	214	A
Superconducting tapes	230	km

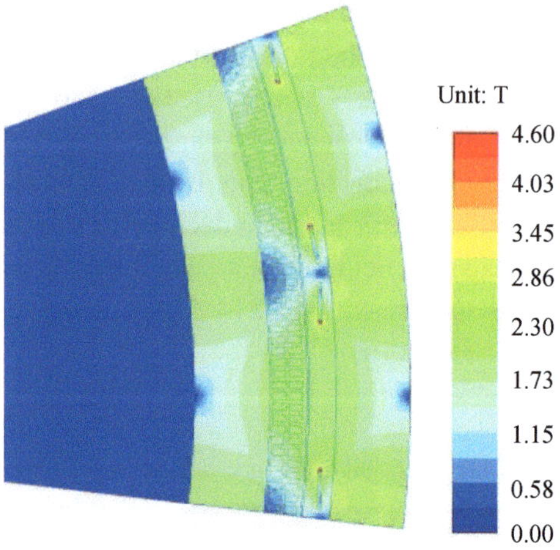

Fig. 7.2: Flux density distribution of the one pole pair model of the superconducting DC generator.

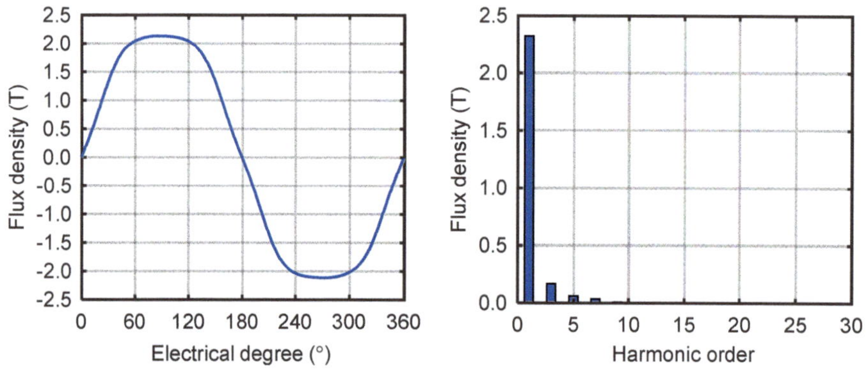

Fig. 7.3: Flux density distribution of the armature winding (in the middle position in radial direction) and its harmonic spectrum of the superconducting DC generator.

To analyze the generator performance, the FE model is coupled with a commutation drive model and the rated load in the simulation is considered as a DC current source with rated current 10 kA. The calculated torque and branch voltage is presented in Fig. 7.4. The peak to peak ripple of the torque and voltage is 0.7% and 1.1%, respectively, which are much smaller compared to the conventional machine. To limit the voltage between the two commutator bars not to exceed a maximum value, typically 30 to 40 V [9], 1176 commutator bars are employed. The coil current and induced voltage is demonstrated in Fig. 7.5. A three-dimensional view of the generator is shown in Fig. 7.6.

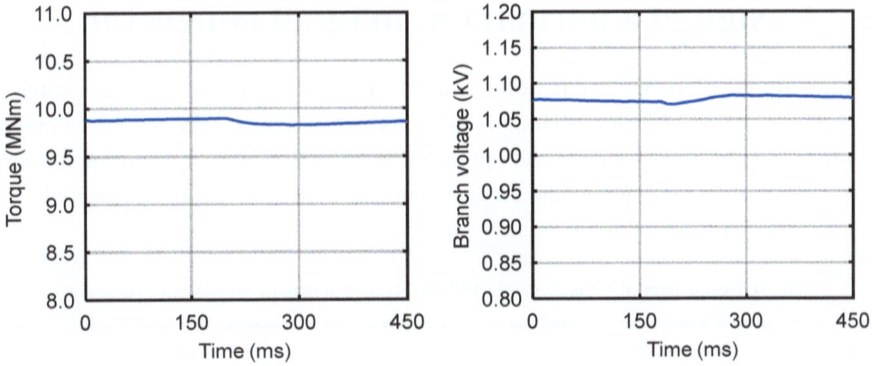

Fig. 7.4: Torque and branch voltage at rated load of the superconducting DC generator.

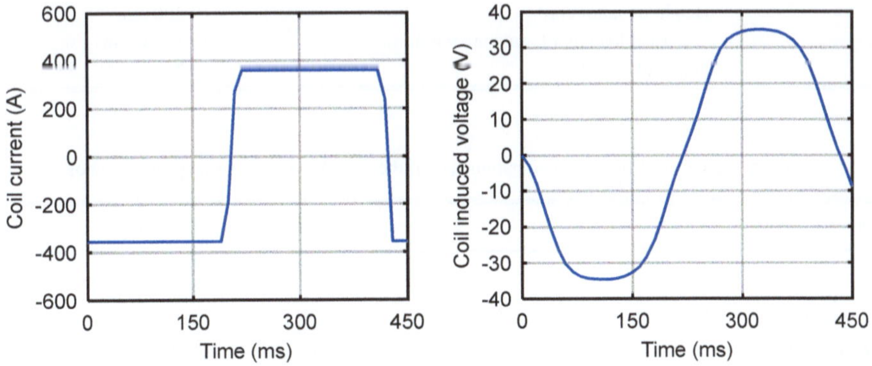

Fig. 7.5: Current and induced voltage of one coil of the superconducting DC generator.

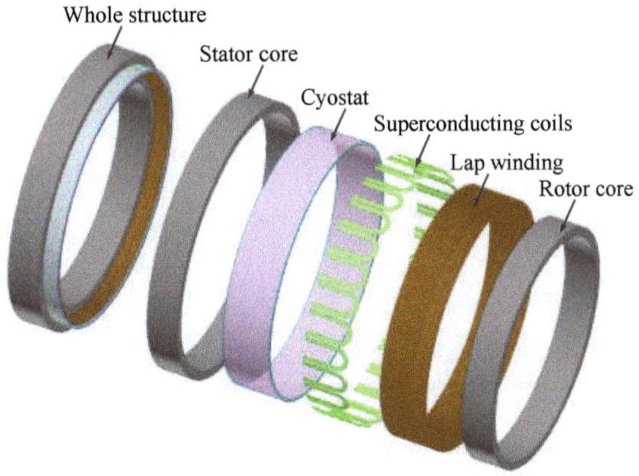

Fig. 7.6: 3D view of the superconducting DC generator (excluding commutator bars and brushes).

7.2 Design of a permanent magnet generator

By combining the electromagnetic design model of the permanent magnet generator, as illustrated in A 3, with mechanical design model in Chapter 4, a 10 MW permanent magnet synchronous generator is designed and optimized in an analytical way. By using the same objective function as the superconducting generator, the optimized design is located.

Table 7.4: Design control parameters of a 10 MW permanent magnet synchronous generator

Symbol	Explanation	Range
D_a	Inner diameter of stator	4-10 m
h_m	Height of the permanent magnets	4-200 mm
h_{ry}	Height of rotor core	10-500 mm
h_{sy}	Height of stator core	10-500 mm
A_a	Electrical loading of armature winding	20-115 kA/m
p	Pole pairs number	10-240

Table 7.5: Design specifications and parameters of the 10 MW permanent magnet generator

Parameter	Value	Unit
Rated power	10	MW
Rated speed	10	1/min
Phase number	3	-
Rated voltage	3.3	kV
Number of poles	240	-
Slot number	1440	-
Stator outer diameter	7370	mm
Stator inner diameter	7267	mm
Stator core thickness	16	mm
Physical air gap	10	mm
Rotor outer diameter	7247	mm
Rotor inner diameter	7165	mm
Rotor core thickness	20	mm
Stack length	2150	mm
Electrical loading	110.3	A/mm
Fundermental flux density	0.81	T
Ratio of pole width to pole pitch	0.7	-
Height of permanent magnet	21.4	mm
PM consuption	5.7	ton

The permanent magnet synchronous generator also employs radial flux structure and the rated speed is 10 (1/min). Surface-mounted PM structure is used, since the fabrication is easier than the interior permanent magnet structure [177]. Three-phase double-layer lap copper windings with iron teeth are adopted in the stator. To make a fair comparison, the air gap length is the same as the superconducting DC generator, which is 10 mm. The rated voltage of the permanent magnet generator is 3.3 kV [97]. The permanent magnets employed are NdFe35 with a remanent flux of 1.2 T. The ranges of the six control parameters are listed in Table 7.4. The constraints of permanent consumption, mass, volume, and efficiency are 1-40 ton, 60-400 ton, 20-200 m^3, 90-99%, respectively.

To investigate the electro-magnetic performance of the permanent magnet generator, a 2D FE model is setup. The design specifications are summarized in Table 7.5. The no-load flux density distribution of one pole model is shown in Fig. 7.7. The flux density of the air gap and its harmonic spectrum are shown in Fig. 7.8. The induced voltage at no-load and electromagnetic torque at rated load are drawn in Fig. 7.9. The peak to peak ripple of the torque is 1.3%, which is higher compared to the superconducting DC generator. A 3D view of the permanent magnet generator is shown in Fig. 7.10.

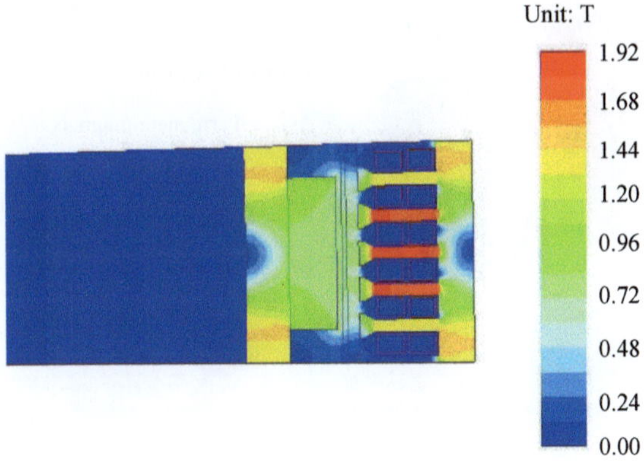

Fig. 7.7: Flux density distribution of the one pole model of the PM generator.

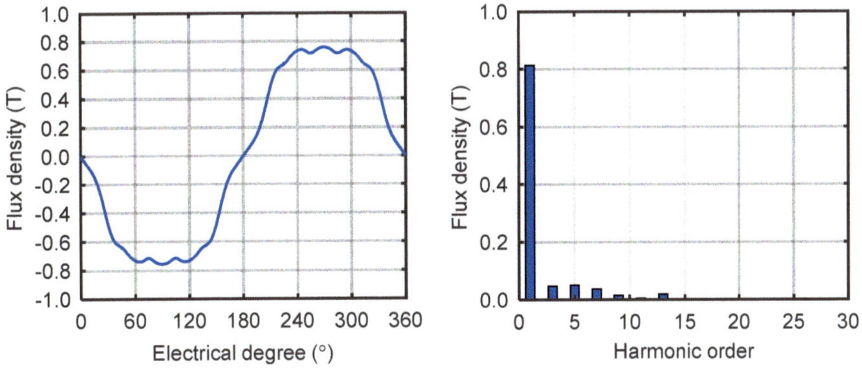

Fig. 7.8: Flux density distribution of the air gap and its harmonic spectrum of the PM generator.

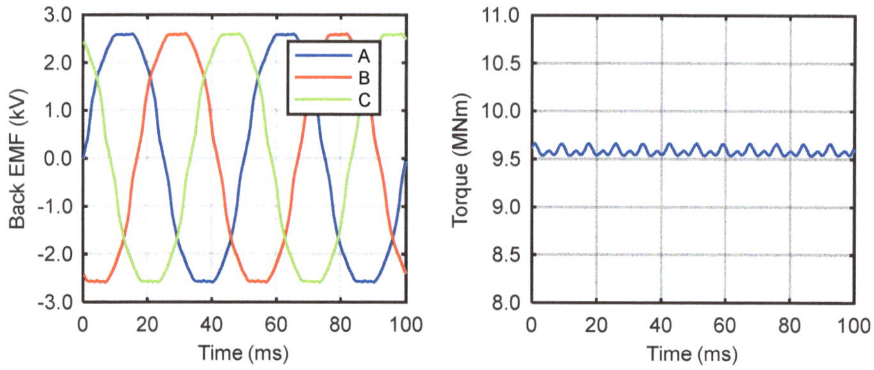

Fig. 7.9: No-load induced voltage and rated torque of the PM generator.

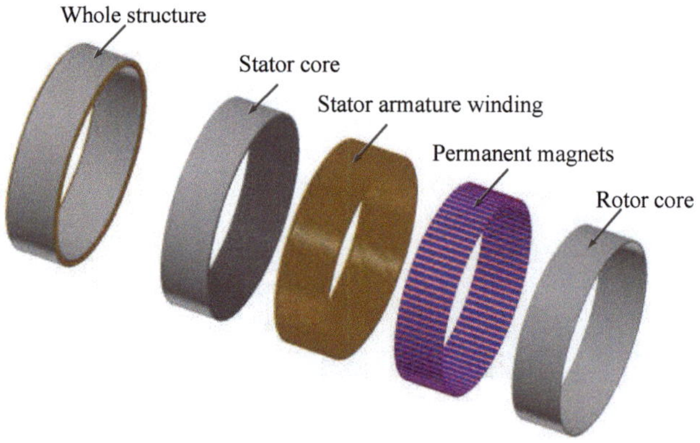

Fig. 7.10: Three-dimensional view of the PM generator.

7.3 Design of a copper coil excited generator

Applying equations in A 4 to the optimization method, the copper coil excited synchronous generator is optimized. This generator also employs radial flux structure three-phase double-layer lap copper windings with iron teeth in the stator. To make a fair comparison, the air gap length is the same as the other two generators. The rated voltage is 3.3 kV, and the ranges of the six control parameters are listed in Table 7.4, same as the permanent magnet generator. Constraints of the copper consumption, mass, volume, and efficiency are 1-40 ton, 60-400 ton, 20-200 m^3, 90-99%. The optimal design specifications are summarized in Table 7.6 and validated by the FE model. The Fig. 7.11 illustrates the no-load flux density distribution of one pole model. The flux density of the air gap is shown in Fig. 7.8, as well as the harmonic spectrum. The induced voltage at no-load and electromagnetic torque at rated load is drawn in Fig. 7.13. The torque peak to peak ripple is 0.8%, which is lower compared to the permanent magnet generator, as its slot number per pole and phase is higher. A 3D view of the permanent magnet generator is shown in Fig. 7.14.

Table 7.6: Design specifications and parameters of the 10 MW copper coil excited generator

Parameter	Value	Unit
Rated power	10	MW
Rated speed	10	1/min
Phase number	3	-
Rated voltage	3.3	kV
Number of poles	96	-
Slot number	1440	-
Stator outer diameter	7224	mm
Stator inner diameter	7044	mm
Stator core thickness	54	mm
Physical air gap	10	mm
Rotor outer diameter	7024	mm
Rotor inner diameter	6672	mm
Rotor core thickness	68	mm
Rotor pole height	89	mm
Rotor pole width	115	mm
Pole shoe height	20	mm
Pole shoe width	161	mm
Stack length	1850	mm
Electrical loading	113.9	A/mm
Fundermental flux density	1.02	T
Ratio of pole width to pole pitch	0.7	-
Copper consumption	31.8	ton

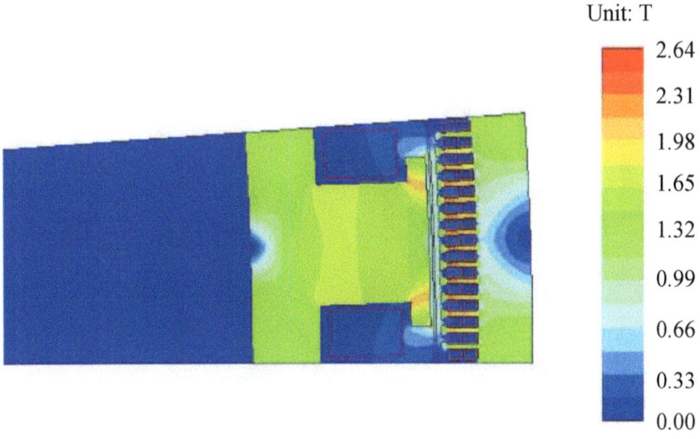

Fig. 7.11: Flux density distribution of the one pole model of the copper coil excited synchronous generator.

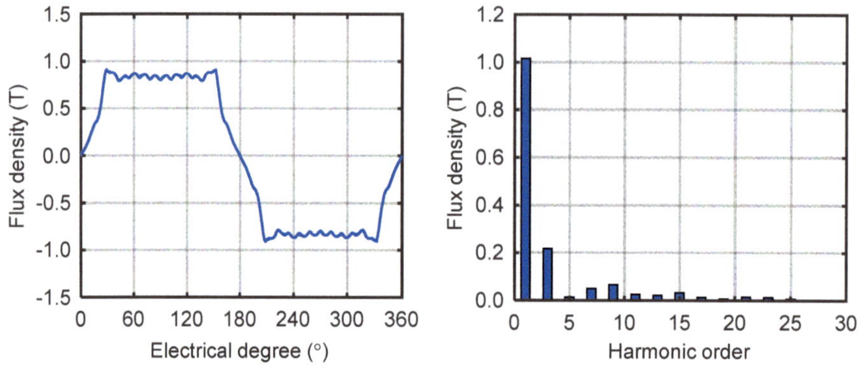

Fig. 7.12: Flux density distribution of the air gap and its harmonic spectrum of the copper coil excited generator.

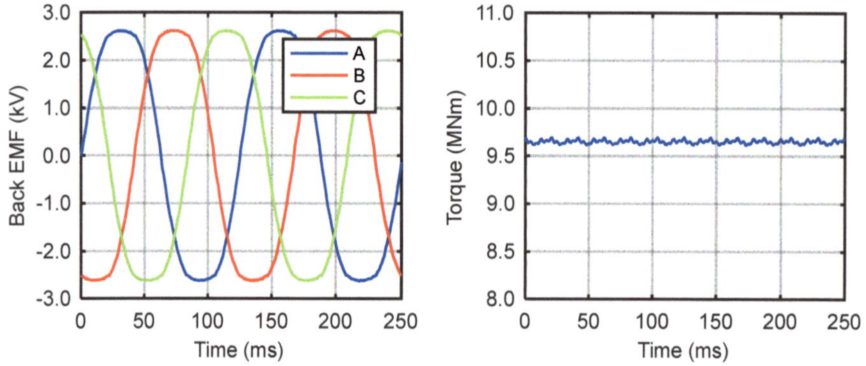

Fig. 7.13: Induced voltage at no load and torque at rated load of the copper coil excited generator.

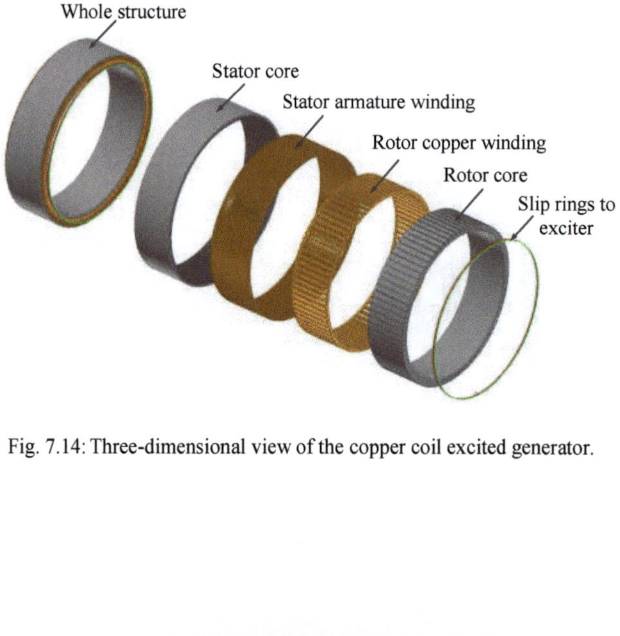

Fig. 7.14: Three-dimensional view of the copper coil excited generator.

7.4 Comparison of the three generators

The torque to volume ratio of the superconducting DC, permanent magnet and copper coil excited synchronous generators are 133 kNm/m^3, 84 kNm/m^3, 92 kNm/m^3, respectively. It is obvious that the torque density of the superconducting DC generator is much higher than the PM and copper coil excited generator. The reason lies in the higher flux density generated by the superconductor. It should be noted that the torque density discussed in this section excludes the outer-casing and air cooling system. The peak to peak torque ripples of the superconducting DC, permanent magnet and copper coil excited synchronous generators are 0.7%, 1.3%, and 0.8%, respectively. The superconducting DC generator has the lowest torque ripple and the permanent magnet generator has the highest. One reason is that in the SC generators the effective electromagnetic air gap is much larger, the armature reaction is relatively small [178]. Moreover, there are no iron teeth in the stator. Correspondingly, the air gap flux density is more sinusoidal and torque ripple is small, as illustrated in Fig. 7.3, Fig. 7.8, Fig. 7.13. One way to decrease the torque ripple of conventional generators is to increase the slot number per pole and phase. However, the tooth width will be reduced. As iron teeth act as mechanical support for the armature winding, there is a upper limit to that number. Furthermore, too many slots also increase the insulation and decrease the slot filling factor. Consequently, copper losses are increased. Another way to decrease cogging torque is to skew stator slots or permanent magnets, making the manufacturing process more complicated. In addition, the drop of the average torque and torque density is another drawback.

The component masses of the three generators are listed in Fig. 7.15, when the inactive materials of the generators are calculated based on rated torque and 3 times rated torque, respectively. The mass of the inactive materials excludes shaft and bearings. To have enough safety margins, the structural supports are calculated according to 3 times the rated torque [61]. The torque to weight ratio of the three generators is 48.6 Nm/kg, 40.8 Nm/kg, and 36.1 Nm/kg, respectively. As no weight reduction method is employed for the structural materials, the masses of the inactive materials are heavier compared to the data in reference [101],[179], to which the structural materials with rated torque calculation are similar without weight reduction method. In other words, the structural supports calculated with rated torque can be used to estimate the total mass with effective weight reduction method as introduced in [54],[180]. It can be seen from Fig. 7.15 that the total weight of the superconducting DC generator is lighter than the permanent magnet and copper coil excited generator. The dominant mass of the superconducting DC generator is the mass of the stator and rotor iron cores, while the mass of the inactive materials dominates in the permanent magnet and copper coil excited generator, if lightweight structure is employed. In this case, torque to weight ratio of the three generators is 78.5 Nm/kg, 61.3 Nm/kg, and 50.4 Nm/kg, respectively. To further reduce the total masses of the superconducting DC generator, the electrical loading can be increased and water cooling of the armature winding is needed instead of air cooling.

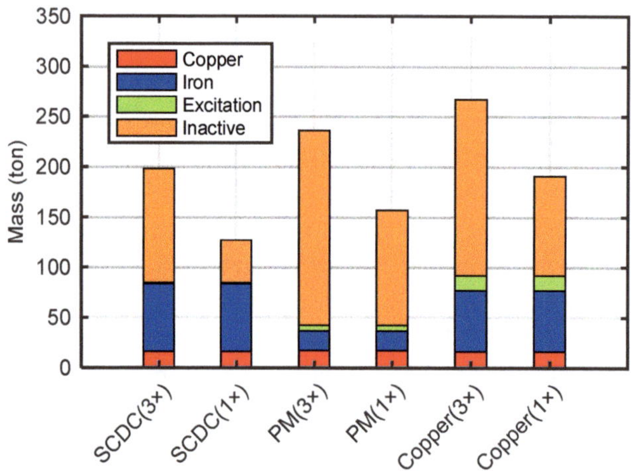

Fig. 7.15: Component masses of the three generators when the inactive materials of the generators are calculated based on rated torque and 3 times rated torque, respectively.

Table 7.7: Component losses of the three generators

Parameter	SC DC	PM	Copper
Copper winding	240 kW	367 kW	347 kW
Iron losses	13 kW	18 kW	13 kW
Copper field winding	--	--	312 kW
Permanent magnets	--	0.2 kW	--
Refrigerator consumption	27 kW	--	--
Brushes losses	24 kW	--	--
Additional losses	100 kW	20 kW	20 kW
Total	404 kW	405 kW	692 kW

The losses of the three generators at rated wind speed are listed in Table 7.7. According to [181], we can assume the losses in the cryostat to be 296 W and the coefficient of performance is assumed to be 1.06% [182]. Then the refrigerator power consumption is 27 kW. To decrease the losses of the permanent magnets, the permanent magnets are made into segments. The efficiency of the superconducting DC, permanent magnet and copper coil excited generator is 96.1%, 96.1% and 93.5%, respectively. The efficiency of the superconducting DC and permanent magnet generator is much higher than the copper coil excited generator, due to the high losses of the copper field winding. Compared to the permanent magnet generator, the superconducting DC generator almost has no advantages in efficiency, due to the high additional losses, brush losses and refrigerator loss. However, the estimation of the additional losses is calculated based on the worst case of the conventional DC machines, which has no compensation winding (if compensation winding is used, the additional losses are halved), whereas the armature reaction to distort the flux density distribution in the superconducting generator is smaller than conventional DC machines, as larger effective air gap is caused by the air gap armature winding and cryostat walls. As a result, the additional losses may be lower than the calculated value. Moreover, smaller resistance brushes can lead to lower brush losses. Consequently, the efficiency of the superconducting DC generator can be higher than the value calculated in Table 7.7.

7.5 Discussion of superconductor consumption

To realize a 10 MW superconducting DC wind generator, the cost of the generator is a big challenge. It has been studied that the levelized cost of energy (LCOE) is linear related to the generator cost [183], and the cost of the superconducting tapes is the main portion of the generator cost. The usage of the superconducting tapes can be minimized by changing the generator design. Fig. 7.16(a) depicts the relation between the amount of the superconducting tapes and outer rotor diameter, while the other parameters such as the flux density in the airgap winding, pole pair number, and electrical loading are the same as the optimal design. By increasing the outer rotor diameter from 4 m to 6 m, the length of the 4 mm width superconducting tapes can be decreased by 150 km. One reason is that the effective length of the generator is reduced by increasing the outer rotor diameter. Another reason is that the maximum parallel and perpendicular flux density in the superconducting coils is also decreased, when the outer rotor diameter is increased. Then the operating current of the superconducting coils is also increased and the number of turns per coil is decreased. As a result, the length of the superconducting tapes is decreased. However, when the outer diameter is over 6 m, the usage of the superconducting tapes is increased, instead of decreased, with the larger outer diameter. Because the pole number is the same, the larger the outer rotor diameter, the longer the ending of the superconducting field coils. In this case, the merits gained by the larger diameter are deteriorated by the increased ending winding length. It is notable that the mass of the generator, including structural and active materials, volume and losses are also increased with the outer rotor diameter, as shown in Fig. 7.16.

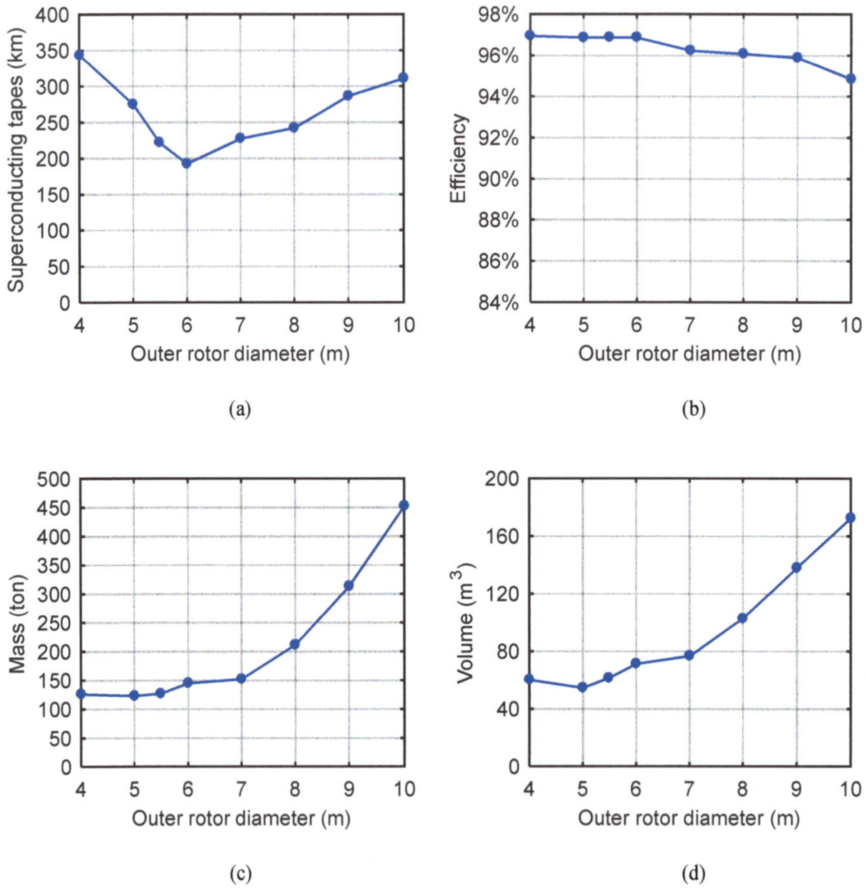

Fig. 7.16: Superconducting tape length, efficiency, mass and volume as a function of outer rotor diameter.

The influence of the pole pair number on the superconducting tape length in case of the other parameters are kept the same as the optimal design is illustrated in Fig. 7.17(a). By decreasing the pole pair number, the superconducting tape length is reduced sharply. When the pole pair number increases, pole pitch is decreased and if the magnetomotive force (MMF) is constant of a superconducting coil, the radial flux density in the air gap is smaller. In order to maintain the same radial flux density in the air gap when pole number changes, the MMF per coil should be enlarged. In addition, more MMF will lead to higher flux density in the superconducting coil and the operating current of the superconducting coil will be decreased, which means more turns are required per coil. Hence, the superconducting tape length is increased greatly when pole number increases. One the contrary, the losses are reduced by increasing pole pair number as shown in Fig. 7.17. The mass and volume of the generator is initially decreased and then increased.

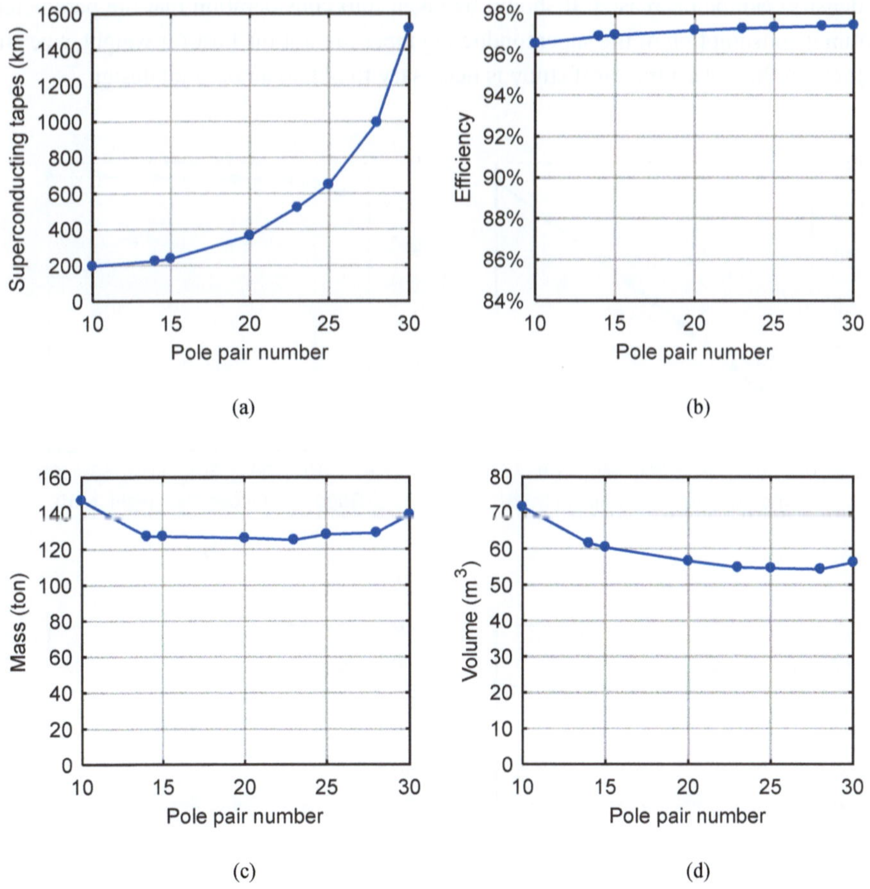

Fig. 7.17: Superconducting tape length, efficiency, mass and volume as a function of pole pair number.

Fig. 7.18 describes the influence of the superconducting coil height on the length of the superconducting tapes, mass, volume and efficiency. With higher superconducting coil, the radial flux density in the air gap is larger. The weight, volume and losses slowly decreased with larger height of superconducting coils, while the superconducting tapes increase sharply. The main reason is that higher superconducting coils not only result in higher flux density in the superconducting coil but also larger effective electromagnetic air gap. Consequently, more MMF per coil is required to achieve the air gap flux density needed. As a result, the superconducting tapes decrease sharply with decreased coil height, that is, smaller flux density.

In summary, to decrease the usage of superconducting tapes, the outer rotor diameter can be increased (to some extent), the pole pair number and the superconducting coil height (flux density) can be decreased. It should be taken into consideration that the parameters for optimal consumption of the superconducting tapes are not the best for weight, size and efficiency. In this case a trade-off study is necessary to obtain an optimal design.

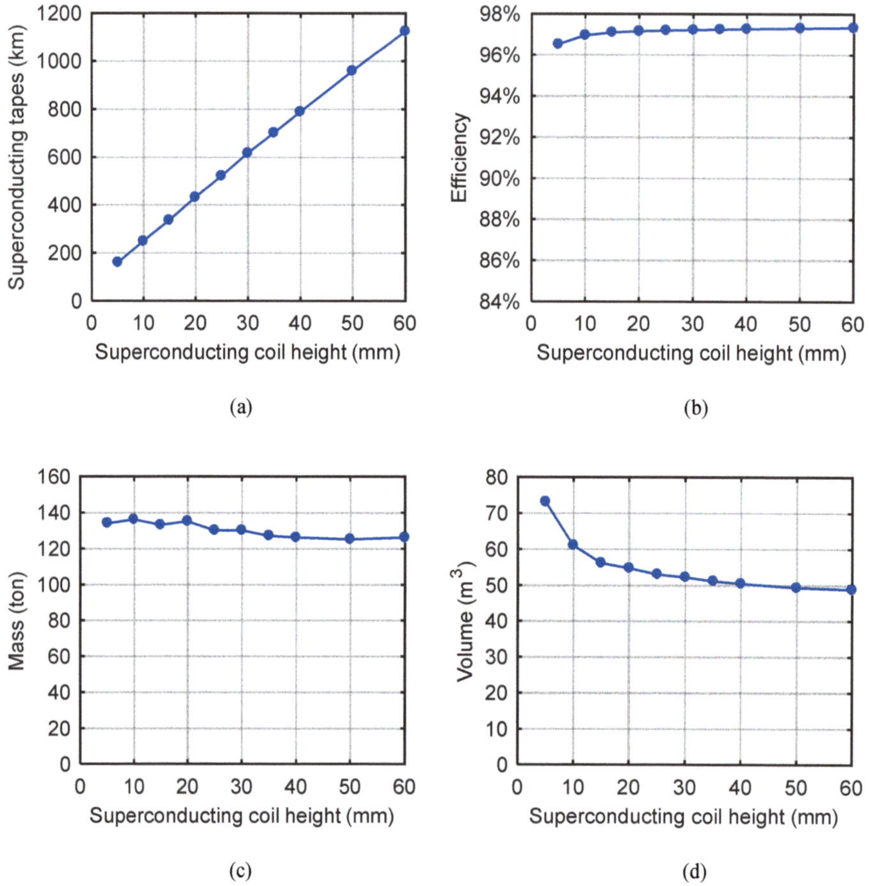

Fig. 7.18: Superconducting tape length, efficiency, mass and volume as a function of superconducting height.

8 Conclusion and outlook

Rapid increase in wind energy asks for new technical solutions in energy conversion system. One key component of the conversion system is the generator which is required to produce higher power with lower cost, lighter weight, smaller size, and higher reliability. In conventional generators, the maximum power is limited by the circumferential current per unit length and the air gap flux density. In order increase the parameters, superconductors are proposed in wind generators. As the superconductor has no DC loss and a high current carrying ability, superconducting generators have advantages in torque density, weight and size. It is found that the average torque to mass ratio of the superconducting generators is 64.5 Nm/kg, which is about double that of permanent magnet generators and tripple that of copper coil excited generators. The average torque to volume ratio of the superconducting generator is 222.2 kNm/m^3, which is around five times that of the permanent magnet generators. Suprapower project designed a 10 MW superconducting generator and successfully built a scaled down demonstrator to validate the technology. Ecoswing project will operate the first superconducting wind turbine at a real wind tower in the near future. It is also predicted that the price of the superconductor will drop to be below 25 $/kA-m in 2030 and with this price, the cost of the superconducting generator can be comparable to the permanent magnet generator. It is believed that more large-scale superconducting wind generators will be seen in a long run.

This work proposes and investigates superconducting DC generators as a way of power generation for wind energy. The main advantages of superconducting DC generators and their integration into the power grid are: high efficiency of energy generation and transportatation and high reliability.

To design a superconducting DC generator all necessary equations were derived and formulated to connect the geometry, properties of the superconducting tapes and iron materials with electrical parameters. The equations were integrated in a design process to automatically find the optimal design with main variables. The design method of the superconducting DC generator can be summarized into five phases: input parameters with control variables and generator specifications, electromagnetic design model for the electrical parameters, mechanical design for structural materials, optimization for locating optimal design by evaluating superconducting tape length, mass, volume, and efficiency, and output parameters with all design parameters of the optimal generator. By this analytical design, a good computational efficiency was obtained.

To accurately calculate the electrical parameters of the superconducting DC generator, as well as to determine the operating current in the superconducting coils, magnetic flux density distribution in the generator is essential. An analytical model based on Poisson and Laplace Equation has been developed to calculate the flux density, which was generated by both superconducting coils and airgap armature winding. Furthermore, the model allowed for the nonlinearity of the ferromagnetic materials and operating relative permeability was determined by an iterative algorithm. Compared with finite element results, it is shown the analytical model to calculate the flux density distribution in the superconducting DC generator is appropriate.

Mass and size of the generator influence the cost of transportation, foundation, and installation. When direct drive generator is increased to larger capacity, the weight of structural materials increases sharply, and even become the dominant weight. It is of significance to include the mechanical structure in the optimization process. Hence, an analytical mechanical model was employed and equations derived based on allowable deflection of stator and rotor support. The stresses on the rotor and stator considered in the model were radial stress, tangential stress and gravity. Finite element software was used to validate the analytical equations and results show that the analytical equations can roughly estimate the structural weight with an accuracy of more than 80%.

Basd on the equations and the design method, a 10 kW superconducting DC generator demonstrator was designed to validate this technology. Superconducting tapes from companies, such as Sunam, Superox, Theva and Superpower, were compared for the demonstrator design and finally Theva tape was chosen for the demonstrator design. The demonstrator has a voltage output of 230 V and six superconducting coils were chosen in the stator. The working temperature of the superconducting coils was designed as 30 K and the operating current is 147 A (0.52 of the critical current). The total superconducting tapes used will be 3.2 km. The total losses in room temperature at rated speed were 1008.6 W, 36.7 W at 80 K, 10.5 W at 30 K. A brush shift of 2.4° was used to smooth the commutation.

Before building the demonstrator the experiments on the key components of the demonstrator have been made, such as the superconducting properties of the HTS tape and the magnetic properties of the iron materials. Superconducting tape has been characterized and tested for the suitability for the application, including angular, magnetic field and temperature dependence with a pulsed current supply. As different manufacurers have their own production techniques, it is impossible to use the same parameters and equations to describe angular, magnetic field and temperature of the critical current for the whole field range of 0-5 T, 30-77 K. Four different iron materials were tested in terms of B-H curves and losses at both room and liquid nitrogen temperature. The results show that the permeability of silicon sheets and losses are higher at 77 K than room temperature. Based on the iron loss model it can be concluded that the temperature mainly has an influence on the eddy current and excess coefficients, while the hysteresis coefficient does not change too much. Moreover, a superconducting double pancake racetrack coil is built and tested. The critical current measured in the coil is consistent with the simulation, which is 48.5 A at 77 K self-field. To run the coil without any risk or quench, it is highly recommended to operate this coil under the current defined by the innermost turn of the top coil, which is 42.3 A at 77 K self-field.

To study the feasibility of the superconducting DC generator with large scale wind energy, a 10 MW superconducting DC generator was designed and compared to a permanent magnet and copper coil excited synchronous generator, from the view of torque density, weight, and efficiency. The torque to volume ratio of the superconducting DC, permanent magnet and copper coil excited synchronous generators are 133 kNm/m^3, 84 kNm/m^3, 92 kNm/m^3; the torque to weight ratio of the three generators is 48.6 Nm/kg, 40.8 Nm/kg, and 36.1 Nm/kg; the efficiency of the three generators is 96.1%, 96.1% and 93.5%, respectively. In order to reduce the superconducting tape consumption so as to reduce the cost, the outer rotor diameter can be increased, the pole pair number and the superconducting coil height can be decreased. However, a trade-off study is necessary to obtain an optimal design based on overall design goal.

From the studies in the thesis, it can be seen that superconducting DC wind generators are highly efficient and compact, but more work needs to be done to realize a first demonstrator. The results presented and experience gained in this work can be seen as a solid base for future development. The major prospect in this thesis is to study the feasibility of superconducting DC generator, design the superconducting coil and demonstrator, and preliminary tests on the key components in the demonstrator. Further investigation will be on the manufacturing process and test of the demonstrator. Emphasis should hereby be given on the cryostat design and manufacture, winding technique of the superconducting coil, thermal, mechanical and electrical tests of the superconducting coil. Detailed quench behavior of the superconducting coil will be performed. Quench model which will provide theoretical analysis of quench and build the basis for quench protection during a fault in the generator will be further investigated. Besides, future studies will be also on how to connect the DC superconducting generator to the grid, what the whole system looks like, and what advantages of the system can have. In this case, power hardware-in-the-loop system could be used to further investigation on the operational behavior of the superconducting DC demonstrator.

Appendix

A.1 Analytic hierarchy process (AHP) to determine the weighting factors

To determine the weighting factors in the objective function (2.31), analytic hierarchy process (AHP) is employed. AHP is a mathematics-based and psychology-based technique for organizing and analysing complex decisions, which was proposed by Thomas L. Saaty in the 1970s. The hierarchy of the problem is shown as in Fig. A.1.1.

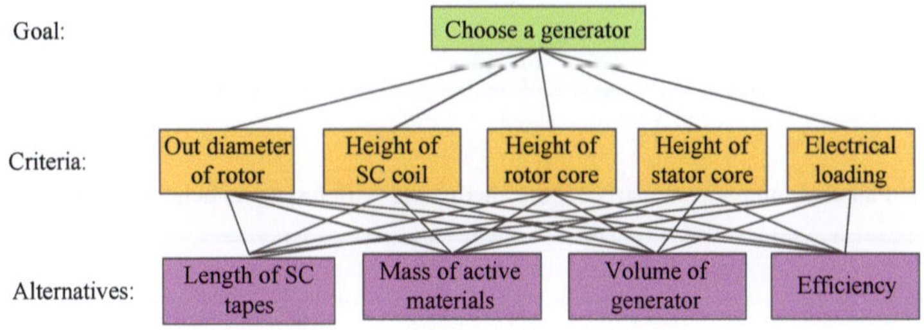

Fig. A.1.1: Hierarchy to determine the weighting factors of the objective function.

To determine the weighting factors in the objective function, the judgement matrix of the alternatives to each criterion is decided by pairwise comparisons based on the importance [184]. The judgement matrix for each criterion is shown from Table A.1.1 to Table A.1.5, respectively. To derive the weighting factors from alternatives to the criteria,equation (A.1.1) is used. The derived weighting factors from alternatives to the criteria are summarized as Table A.1.6. The judgement of matrix from the criteria to the goal is listed in Table A.1.7. Its derived weighting factors and constancy are in Table A.1.8. Hence, the final normalized weighting factors w_{tl}, w_{tm}, w_{tv}, and w_e from the alternatives to the goal are calculated by synthesising Table A.1.6 and Table A.1.8, as shown in equation (A.1.2), which is underlined in Table A.1.9.

$$\omega_i = \frac{1}{n}\sum_{j=1}^{n}\frac{a_{ij}}{\sum_{k=1}^{n} a_{kj}} \qquad i = 1, 2, \cdots, n \tag{A.1.1}$$

where, a_{ij} is the element of the pairwise comparison matrix in i row j column, n is the matrix order, and w_i is the weighting factor from i alternatives with respect to a criterion.

$$W_g = W_a W_c \tag{A.1.2}$$

where, W_g is the weighting factor matrix from alternatives to the goal, W_a is the weighting factor matrix from alternatives to the criteria, and W_c is the weighting factor matrix from criteria to the goal.

Table A.1.1: Pairwise comparison matrix for the alternatives with respect to outer diameter of rotor

	Length	Mass	Volume	Efficiency
Length	1	3	3	5
Mass	1/3	1	1	3
Volume	1/3	1	1	3
Efficiency	1/5	1/3	1/3	1

Table A.1.2: Pairwise comparison matrix for the alternatives with respect to height of superconducting coil

	Length	Mass	Volume	Efficiency
Length	1	3	3	5
Mass	1/3	1	1	3
Volume	1/3	1	1	3
Efficiency	1/5	1/3	1/3	1

Table A.1.3: Pairwise comparison matrix for the alternatives with respect to height of rotor core

	Length	Mass	Volume	Efficiency
Length	1	1/3	1/3	3
Mass	3	1	1	3
Volume	3	1	1	3
Efficiency	1/3	1/3	1/3	1

Table A.1.4: Pairwise comparison matrix for the alternatives with respect to height of stator core

	Length	Mass	Volume	Efficiency
Length	1	1/3	1/3	3
Mass	3	1	1	3
Volume	3	1	1	3
Efficiency	1/3	1/3	1/3	1

Table A.1.5: Pairwise comparison matrix for the alternatives with respect to electrical loading

	Length	Mass	Volume	Efficiency
Length	1	5	5	3
Mass	1/5	1	1	1/3
Volume	1/5	1	1	1/3
Efficiency	1/3	3	3	1

Table A.1.6: Derived weighting factors from alternatives to the criteria

	Outer diameter	SC coil height	Rotor coil height	Stator coil height	Electrical loading
Length	0.5222	0.5222	0.1686	0.1686	0.5595
Mass	0.1998	0.1998	0.3679	0.3679	0.0955
Volume	0.1998	0.1998	0.3679	0.3679	0.0955
Efficiency	0.0781	0.0781	0.0956	0.0956	0.2495
Consistancy	0.0161	0.0161	0.0572	0.0572	0.0161

Table A.1.7: Pairwise comparison matrix of the criteria with respect to the goal

	Outer diameter	SC coil height	Rotor coil height	Stator coil height	Electrical loading
Outer diameter	1	3	8	8	3
SC coil height	1/3	1	6	6	1
Rotor coil height	1/8	1/6	1	1	1/6
Stator coil height	1/8	1/6	1	1	1/6
Electrical loading	1/3	1	6	6	1

Table A.1.8: Derived weighting factors from the criteria to the goal

	Outer diameter	SC coil height	Rotor coil height	Stator coil height	Electrical loading	Consistency
Goal	0.4821	0.2163	0.0426	0.0426	0.2163	0.0237

Table A.1.9: Derived weighting factors from alternatives to the goal

	Length w_{tl}	Mass w_{tm}	Volume w_{tv}	Efficiency w_e
Goal	0.5001	0.1915	0.1915	0.1167

A.2 Equations to calculate flux density

Table A.2.1: Equations to calculate flux density with stator iron core

Coeff.	*np≠2*	*np=2*
E	$r_{si}^{2np}[1-\left(\frac{\mu_{rs}-1}{\mu_{rs}+1}\right)^2(\frac{r_{si}}{r_{so}})^{2np}] \Big/ \left\{\left(\frac{\mu_{rs}-1}{\mu_{rs}+1}\right)[1-(\frac{r_{si}}{r_{so}})^{2np}]\right\}$	$r_{si}^{4}[1-\left(\frac{\mu_{rs}-1}{\mu_{rs}+1}\right)^2(\frac{r_{si}}{r_{so}})^{4}] \Big/ \left\{\left(\frac{\mu_{rs}-1}{\mu_{rs}+1}\right)[1-(\frac{r_{si}}{r_{so}})^{4}]\right\}$
C_{n3}	$-\frac{k_e r_{sci}^{np+2}}{2np(2+np)}$	$-\frac{k_e r_{sci}^4}{16}$
C_{n4}	$-\frac{k_e (r_{sci}^{np+2}-r_{sco}^{np+2})}{2np(2+np)}$	$-\frac{k_e (r_{sci}^4-r_{sco}^4)}{16}$
A_{n4}	$\frac{1}{E}C_{n4}$	$\frac{1}{E}C_{n4}$
A_{n2}	$A_{n4}-\frac{k_e (r_{sci}^{2-np}-r_{sco}^{2-np})}{2np(2-np)}$	$A_{n4}-\frac{k_e (\ln r_{sci}-\ln r_{sco})}{4}$
A_{n3}	$A_{n4}+\frac{k_e r_{sco}^{2-np}}{2np(2-np)}$	$A_{n4}+\frac{k_e \ln r_{sco}}{4}+\frac{k_e}{16}$
C_{n5}	$\frac{(1-\mu_{rs})A_{n4}r_{si}^{2np}+(\mu_{rs}+1)C_{n4}}{2}$	$\frac{(1-\mu_{rs})A_{n4}r_{si}^{4}+(\mu_{rs}+1)C_{n4}}{2}$
A_{n5}	$\frac{(\mu_{rs}+1)A_{n4}+(1-\mu_{rs})C_{n4}r_{si}^{-2np}}{2}$	$\frac{(\mu_{rs}+1)A_{n4}+(1-\mu_{rs})C_{n4}r_{si}^{-4}}{2}$
C_{n7}	$\frac{2C_{n5}}{\mu_{rs}+1}$	$\frac{2C_{n5}}{\mu_{rs}+1}$
A_{n1}	A_{n2}	A_{n2}
A_{n6}	A_{n1}	A_{n1}
C_{n1}	0	0
C_{n2}	0	0
C_{n6}	0	0
A_{n7}	0	0

Table A.2.2: Equations to calculate flux density with rotor iron core

Coeff.	*np≠2*	*np=2*
K	$r_{ro}^{2np}\left(\frac{\mu_{rr}-1}{\mu_{rr}+1}\right)[1-(\frac{r_{ri}}{r_{ro}})^{2np}] \Big/ \left[1-\left(\frac{\mu_{rr}-1}{\mu_{rr}+1}\right)^2(\frac{r_{ri}}{r_{ro}})^{2np}\right]$	$r_{ro}^{4}\left(\frac{\mu_{rr}-1}{\mu_{rr}+1}\right)[1-(\frac{r_{ri}}{r_{ro}})^{4}] \Big/ \left[1-\left(\frac{\mu_{rr}-1}{\mu_{rr}+1}\right)^2(\frac{r_{ri}}{r_{ro}})^{4}\right]$
A_{n3}	$\frac{k_e r_{sco}^{2-np}}{2np(2-np)}$	$\frac{k_e \ln r_{sco}}{4}+\frac{k_e}{16}$

A_{n2}	$-\frac{k_e(r_{sci}^{2-np}-r_{sco}^{2-np})}{2np(2-np)}$	$-\frac{k_e(\ln r_{sci}-\ln r_{sco})}{4}$
C_{n2}	KA_{n2}	KA_{n2}
C_{n3}	$C_{n2}-\frac{k_e r_{sci}^{np+2}}{2np(2+np)}$	$C_{n2}-\frac{k_e r_{sci}^{4}}{16}$
C_{n4}	$C_{n2}-\frac{k_e(r_{sci}^{np+2}-r_{sco}^{np+2})}{2np(2+np)}$	$C_{n2}-\frac{k_e(r_{sci}^{4}-r_{sco}^{4})}{16}$
A_{n1}	$\frac{(\mu_{rr}+1)A_{n2}+(1-\mu_{rr})C_{n2}r_{ro}^{-2np}}{2}$	$\frac{(\mu_{rr}+1)A_{n2}+(1-\mu_{rr})C_{n2}r_{ro}^{-4}}{2}$
C_{n1}	$\frac{(1-\mu_{rr})A_{n2}r_{ro}^{2np}+(\mu_{rr}+1)C_{n2}}{2}$	$\frac{(1-\mu_{rr})A_{n2}r_{ro}^{4}+(\mu_{rr}+1)C_{n2}}{2}$
A_{n6}	$\frac{2A_{n1}}{\mu_{rr}+1}$	$\frac{2A_{n1}}{\mu_{rr}+1}$
C_{n5}	C_{n4}	C_{n4}
C_{n7}	C_{n5}	C_{n5}
A_{n4}	0	0
A_{n5}	0	0
C_{n6}	0	0
A_{n7}	0	0

Table A.2.3: Equations to calculate flux density with stator and rotor iron cores

Coeff.	$np \neq 2$	$np=2$
A_{n3}	$\frac{k_e r_{sco}^{2-np}}{2np(2-np)}$	$\frac{k_e \ln r_{sco}}{4}+\frac{k_e}{16}$
A_{n2}	$-\frac{k_e(r_{sci}^{2-np}-r_{sco}^{2-np})}{2np(2-np)}$	$-\frac{k_e(\ln r_{sci}-\ln r_{sco})}{4}$
C_{n3}	$-\frac{k_e r_{sci}^{np+2}}{2np(2+np)}$	$-\frac{k_e r_{sci}^{4}}{16}$
C_{n4}	$-\frac{k_e(r_{sci}^{np+2}-r_{sco}^{np+2})}{2np(2+np)}$	$-\frac{k_e(r_{sci}^{4}-r_{sco}^{4})}{16}$
A_{n1}	A_{n2}	A_{n2}
A_{n6}	A_{n1}	A_{n1}
C_{n5}	C_{n4}	C_{n4}
C_{n7}	C_{n5}	C_{n5}
A_{n4}	0	0
A_{n5}	0	0
C_{n1}	0	0
C_{n2}	0	0
C_{n6}	0	0
A_{n7}	0	0

A.3 Equations to design permanent magnet synchronous generator

If the saturation of the ferromagnetic material is neglected, the magnetic flux of a pole is expressed as

$$\phi = B_g \frac{\pi D_\delta l_s}{2p} = \left(\frac{2}{\pi}\widehat{B}_g\right)\frac{\pi D_\delta l_s}{2p} \tag{A.3.1}$$

where D_δ is the air gap diameter, p is the pole pair number, l_s is the stack length of the stator, B_g is the average value of the magnetic flux density in the airgap, $\widehat{B}_g$ is the amplitude of the flux density.

By employing carter factor K_c , the slot opening can be taken into consideration.

$$K_c = \frac{\tau_s}{\tau_s - \kappa(\tau_s - b_t)} \tag{A.3.2}$$

$$\kappa = \frac{2}{\pi}\left[\tan^{-1}\left(\frac{\tau_s - b_t}{2\delta}\right) - \left(\frac{2\delta}{\tau_s - b_t}\right)\ln\sqrt{1+\left(\frac{\tau_s - b_t}{2\delta}\right)^2}\right] \tag{A.3.3}$$

where δ is air gap length, κ_c is the Carter factor, τ_s is the slot pitch, b_t is teeth width.

The magnetic flux density above a magnet in the air gap $\widehat{B}_{g,r}$ is

$$\widehat{B}_{g,r} = \frac{\dfrac{h_m B_{rm}}{u_{rm}}}{\dfrac{h_m}{u_{rm}} + \kappa_c \delta} = \frac{h_m B_{rm}}{u_{rm}\delta_{eff}} \tag{A.3.4}$$

where h_m is the height of the permanent magnet, and δ_{eff} is the effective air gap length, B_{rm} is the remanent flux density, u_{rm} is the relative permeability of the permanent magnet.

By Fourier analysis, the fundamental harmonic amplitude $\widehat{B}_g$ is

$$\widehat{B}_g = \frac{4}{\pi}\frac{h_m B_{rm}}{u_{rm}\delta_{eff}}\sin\left(\frac{b_p \pi}{2\tau_p}\right) \tag{A.3.5}$$

where b_p is the width of the pole, τ_p is the pole pitch.

Assuming the flux density, flux linkage and current linkage of the armature winding are sinusoidal, the multiple-phase winding magnetizing inductance is calculated as [33].

$$L_{sm} = \frac{\mu_0 m D_\delta l_s \left(k_w N_s\right)^2}{\pi \delta_{eff} p^2} \tag{A.3.6}$$

where L_{sm} is the magnetizing inductance, N_s is the total number of the stator turns per phase, m is the phase number, k_w is the winding factor, which can be calculated as

$$k_w = k_d k_p = \frac{\sin(\frac{\pi}{2m})}{q \sin\left(\frac{\pi}{2mq}\right)} \sin\left(\frac{W}{\tau_p}\frac{\pi}{2}\right) \tag{A.3.7}$$

where k_d is the distribution factor, k_p is the pitch factor, q is the number of slot per pole and phase, W is the coil width, τ_p is the pole pitch.

To calculate the d-axis and q-axis inductance, L_{dm} and L_{qm}, it is estimated as [185]

$$L_{dm} = L_{qm} = L_{sm} \tag{A.3.8}$$

The stator leakage inductance is calculated as[33].

$$L_{s\sigma} = L_{sq} + L_\delta + L_u + L_d + L_w \tag{A.3.9}$$

where $L_{s\sigma}$ is the stator leakage inductance, L_{sq} is the skew leakage inductance, L_δ is the air gap leakage inductance, L_u is the slot leakage inductance, L_d is the tooth tip leakage inductance, L_w is the end winding leakage inductance.

The induced voltage by airgap magnetomotive force in a stator winding, E_g is

$$E_g = \sqrt{2}\pi N_s k_w f \phi = \frac{\sqrt{2}}{2} N_s k_w \omega_m l_s D_\delta \widehat{B}_g \tag{A.3.10}$$

where, ω_m is the mechanical angular frequency.

In general phasor diagram, as illustrated in Fig. A.3.1(a), the inner power factor angle is

$$\psi = \arctan \frac{U_a \sin\varphi + I_a X_{s\sigma} + I_a X_{mq}}{U_a \cos\varphi + I_a R_a} \tag{A.3.11}$$

where φ is the power factor angle, which is angle between U_a and I_a; U_a is the terminal phase voltage, I_a is the terminal phase current.

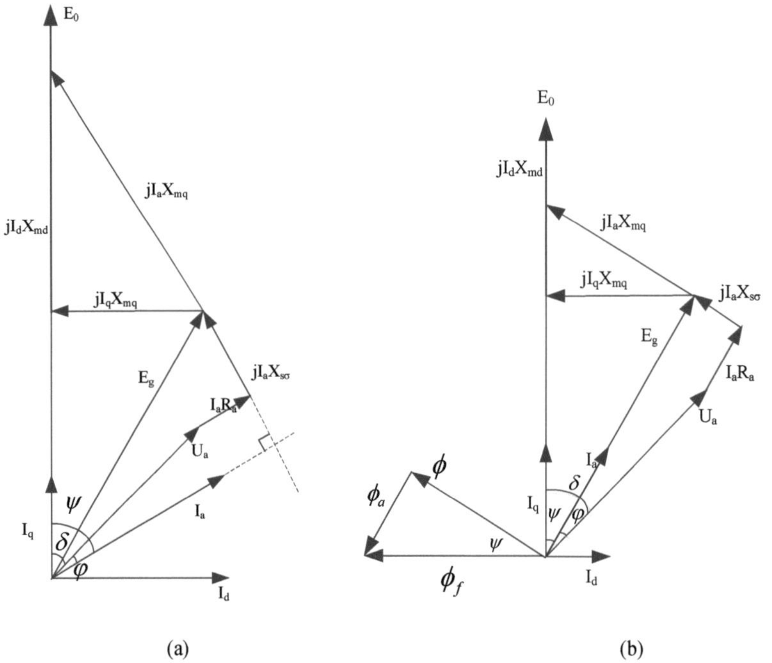

Fig. A.3.1: Phasor diagram, (a) general phasor diagram, and (b) phasor diagram with I_a and E_g in-phase.

The electromagnetic power, P_e, which is the power into the stator is calculated as

$$P_e = mU_a I_a \cos\varphi + mI_a^2 R_a = mI_a (U_a \cos\varphi + I_a R_a) = mE_g I_a \cos\psi_0 \tag{A.3.12}$$

In order to have maximum power into the stator, the I_a and E_g need to be in-phase, that is, ψ_0 is equal to 0.

When I_a and E_g are in-phase, the phasor diagram is shown in Fig. A.3.1(b), the angle between E_0 and I_a, ψ, which is the inner power factor angle is

$$\psi = \arctan\frac{I_a X_{mq}}{U_a \cos\varphi + I_a R_a} = \arctan\frac{I_a X_{mq}}{E_g} \tag{A.3.13}$$

where $U_a \cos\varphi = E_g - I_a R_a$. The output terminal power, P_{out}, and the back-EMF of the no-load condition are

$$P_{out} = mI_aU_a\cos\varphi \tag{A.3.14}$$

$$E_0 = X_{dm}I_d + \sqrt{E_g^2 - \left(X_{qm}I_q\right)^2} = X_{dm}I_a\sin\psi + \sqrt{E_g^2 - \left(X_{qm}I_a\cos\psi\right)^2} \tag{A.3.15}$$

The flux density in the stator iron teeth is calculated as

$$B_t = \frac{1}{2}\left(\widehat{B}_g + \widehat{B}_{g,r}\right)\frac{\tau_s}{b_t} \tag{A.3.16}$$

where B_t is the flux density in the stator iron teeth, τ_s is the slot pitch, b_t is the teeth width. The average value of $\widehat{B}_g$ and $\widehat{B}_{g,r}$ is taken because the amplitude of rectangular flux waveform would underestimate the flux density in the stator iron teeth, while the amplitude of the sinusoidal flux waveform would overestimate the flux density in the iron teeth.

Another way to calculate the flux density in the stator iron teeth is as follow,

$$\varphi_t = \tau_s\widehat{B}_g l_s - (\tau_s - b_t)l_sB_u = B_tb_tl_s \tag{A.3.17}$$

where φ_t is the flux into the stator iron tooth, and B_u is the flux density in the slot area

Hence,

$$B_t = \frac{\tau_s}{b_t}\widehat{B}_g - \frac{\tau_s - b_t}{b_t}B_u \tag{A.3.18}$$

Since the tangential component of the field strength is continuous at the interface of the iron and the air, that is $H_t = H_u$,

Then, the B_t equation (A.3.18) can be rewritten as

$$B_t = \frac{\tau_s}{b_t}\widehat{B}_g - \frac{\tau_s - b_t}{b_t}\mu_0H_t \tag{A.3.19}$$

The crossing point of this B_t in eq. (A.3.19) with *B-H* curve is the operating B_t, H_t, and μ_t.

The flux of one pole, ϕ,in the air gap and the flux density in the stator yoke, B_{sy}, are

$$\phi = B_g\frac{\pi D_\delta l_s}{2p} = \left(\frac{2}{\pi}\widehat{B}_g\right)\frac{\pi D_\delta l_s}{2p} = \left(\frac{2}{\pi}\widehat{B}_g\right)\tau_pl_s \tag{A.3.20}$$

$$B_{sy} = \frac{\phi}{2h_{sy}l_s} = \frac{\widehat{B}_g \tau_p}{\pi h_{sy}} \tag{A.3.21}$$

where, h_{sy} is the height of the stator yoke

From the phasor diagram with I_a and E_g in-phase, as shown in Fig. A.3.1(b), it can be seen that the main flux in the air gap generated by the pole ϕ can be calculated by flux in the air gap at no-load condition, which is ϕ_f.

$$\phi_f = \frac{\phi}{\cos\psi} \tag{A.3.22}$$

Hence, the flux density in the rotor yoke, B_{ry}, is calculated as

$$B_{ry} = \frac{\phi_f}{2h_{ry}l_s} = \frac{\widehat{B}_g \tau_p}{\pi h_{ry}\cos\psi} \tag{A.3.23}$$

where h_{ry} is the height of the rotor yoke.

By using the flux density and *B-H* curve, the relative permeability of rotor core, stator teeth and stator core, μ_{ry}, μ_t, μ_{sy}, and magnetic strength of rotor core, stator teeth and stator core, H_{ry}, H_t, H_{sy} can be calculated. Hence, the total ampere-turns (magnetomotive force) is

$$F_{total} = H_{ry}h_{ry} + H_g\delta_{eff} + H_t b_t + H_{sy}h_{sy} = \cos\psi \frac{B_{rm}h'_{rm}}{\mu_0\mu_{rm}} \tag{A.3.24}$$

By comparing h_m and h'_{rm}, δ_{eff} is iterated to take the saturation into consideration. For example, if $h'_{rm} > h_m$, which means the setting MMF is not enough for the machine as iron cores are highly saturated, and the effective airgap is larger than the value used before. If h_m is equal to h'_{rm}, iteration calculation stops. Otherwise the iteration loop. H_g is the magnetic strength of the air gap and calculated according to the rectangular flux density. Note: $\frac{B_{rm}h_{rm}}{\mu_0\mu_{rm}}$ is the equivalent to the MMF produced by $N_f I_f$.

A.4 Equations to design copper coil excited synchronous generator

The magnetomotive force of one pole pair in the air gap is shown in Fig A.4.1, the Fourier transform of the amplitude of the fundamental component is expressed as

$$\widehat{F}_{airgap1} = \frac{4N_f I_f}{\pi}\sin\left(\frac{b_p \pi}{2\tau_p}\right) \tag{A.4.1}$$

where $F_{airgap1}$ is the average value of the fundamental component of the magnetomotive force of one pole pair in the air gap, $\widehat{F}_{airgap1}$ the amplitude of the fundamental component of the magnetomotive force, N_f is the number of field winding, If is the current in the field winding, b_p is the width of the pole, τ_p is the pole pitch.

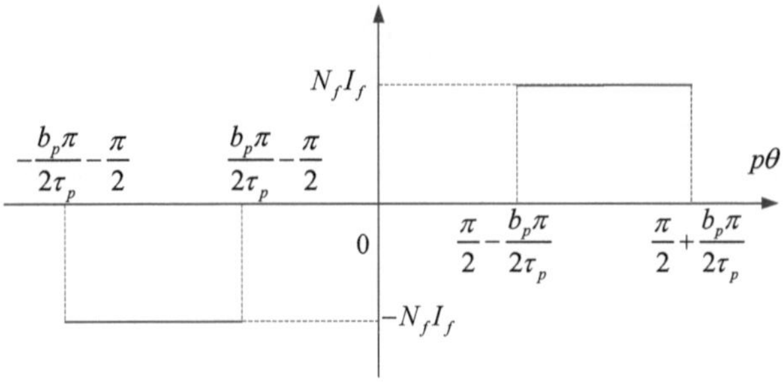

Fig. A.4.1 Magnetomotive force of one pole pair in the air gap.

If the saturation of the ferromagnetic material is neglected, the magnetic flux of a pole is

$$\phi = F_{airgap1} G \tag{A.4.2}$$

where ϕ is the magnetic flux of a pole through the air gap, G is the permeance of the air gap and calculated by

$$G = \frac{\mu_0 \dfrac{\pi D_\delta l_s}{2p}}{\delta_{eff}} \tag{A.4.3}$$

where D_δ is the air gap diameter, p is the pole pair number, l_s is the stack length of the stator, and δ_{eff} is the effective air gap length.

The magnetic flux can be also expressed as a function of the magnetic flux density, as shown in eq.(A.4.4), where B_g is the average value of the magnetic flux density in the airgap, $\widehat{B}_g$ is the amplitude of the flux density.

$$\phi = B_g \frac{\pi D_\delta l_s}{2p} = \left(\frac{2}{\pi}\widehat{B}_g\right)\frac{\pi D_\delta l_s}{2p} \tag{A.4.4}$$

By combining all the equations, it can be derived that the magnetic flux density is

$$\widehat{B}_g = \frac{4\mu_0 N_f I_f}{\pi \delta_{eff}} \sin\left(\frac{b_p \pi}{2\tau_p}\right) \tag{A.4.5}$$

The rectangular flux density (before the Fourier transform) is

$$\widehat{B}_{g,r} = \frac{\mu_0 N_f I_f}{\delta_{eff}} \tag{A.4.6}$$

By employing carter factor the initial effective airgap δ_{eff} can be calculated as

$$\delta_{eff} = \kappa_c \delta \tag{A.4.7}$$

where δ is the air gap length, κ_c is the Carter factor which is

$$\kappa_c = \frac{\tau_s}{\tau_s - \kappa(\tau_s - b_t)} \tag{A.4.8}$$

$$\kappa = \frac{2}{\pi}\left[\tan^{-1}\left(\frac{\tau_s - b_t}{2\delta}\right) - \left(\frac{2\delta}{\tau_s - b_t}\right)\ln\sqrt{1+\left(\frac{\tau_s - b_t}{2\delta}\right)^2}\right] \tag{A.4.9}$$

Assuming the flux density, flux linkage and current linkage of the armature winding are sinusoidal, the multiple-phase winding magnetizing inductance is calculated as

$$L_{sm} = \frac{\mu_0 m D_\delta l_s (k_w N_s)^2}{\pi \delta_{eff} p^2} \tag{A.4.10}$$

where L_{sm} is the magnetizing inductance, k_w is the winding factor, N_s is the total number of the stator turns per phase, m is the phase number[33].The winding factor, k_w, is the product of distribution factor and short pitch factor

$$k_w = k_d k_p = \frac{\sin(\frac{\pi}{2m})}{q\sin\left(\frac{\pi}{2mq}\right)} \sin\left(\frac{W}{\tau_p}\frac{\pi}{2}\right) \tag{A.4.11}$$

where k_d is the distribution factor, k_p is the pitch factor, q is the number of slot per pole and phase, W is the coil width, τ_p is the pole pitch, m is the number of phase.

To calculate the d-axis and q-axis inductance, L_{dm} and L_{qm}, it is estimated as[185]

$$L_{dm} = 0.95 L_{sm} \tag{A.4.12}$$

$$L_{qm} = 0.5 L_{sm} \tag{A.4.13}$$

The stator leakage inductance is calculated as [33].

$$L_{s\sigma} = L_{sq} + L_\delta + L_u + L_d + L_w \tag{A.4.14}$$

where $L_{s\sigma}$ is the stator leakage inductance, L_{sq} is the skew leakage inductance, L_δ is the air gap leakage inductance, L_u is the slot leakage inductance, L_d is the tooth tip leakage inductance, L_w is the end winding leakage inductance.

The maximum value of the flux linkage of the stator winding is

$$\hat{\psi}_s = k_w N_s \frac{\pi D_\delta}{2p} l_s (\frac{2}{\pi}\hat{B}_g) \tag{A.4.15}$$

Then the maximum value of the mutual inductance between stator and rotor is

$$\widehat{M}_{sf} = \frac{\hat{\psi}_s}{I_f} = \frac{4\mu_0 k_w N_s N_f l_s D_\delta}{\pi p \delta_{eff}} \sin\left(\frac{b_p \pi}{2\tau_p}\right) \tag{A.4.16}$$

The induced voltage by airgap magnetomotive force in a stator winding, E_g (rms value) is

$$E_g = \sqrt{2}\pi N_s k_w f \phi = \frac{\sqrt{2}}{2} N_s k_w \omega_m l_s D_\delta \hat{B}_g \tag{A.4.17}$$

where, ω_m is the mechanical angular frequency. In general phasor diagram, as illustrated in Fig. A.4.2(a), the inner power factor angle is

$$\psi = \arctan \frac{U_a \sin\varphi + I_a X_{s\sigma} + I_a X_{mq}}{U_a \cos\varphi + I_a R_a} \tag{A.4.18}$$

where φ is the power factor angle, which is angle between U_a and I_a; U_a is the terminal phase voltage, I_a is the terminal phase current.

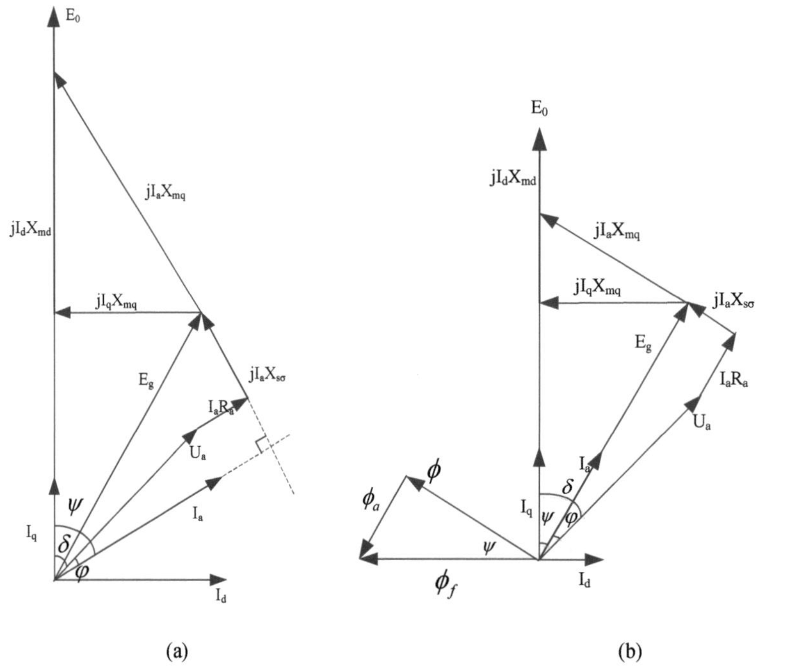

Fig. A.4.2: Phasor diagram, (a) general phasor diagram, and (b) phasor diagram with I_a and E_g in-phase.

The electromagnetic power which is the power into the stator is calculated as

$$P_e = mU_a I_a \cos\varphi + mI_a^2 R_a = mI_a(U_a \cos\varphi + I_a R_a) = mE_g I_a \cos\psi_0 \tag{A.4.19}$$

To have maximum power into the stator, the I_a and E_g need to be in-phase, that is, ψ_0 is equal to 0.

When I_a and E_g are in-phase, the phasor diagram is shown in Fig. A.4.2(b), the angle between E_0 and I_a, ψ , which is the inner power factor angle is

$$\psi = \arctan\frac{I_a X_{mq}}{U_a \cos\varphi + I_a R_a} = \arctan\frac{I_a X_{mq}}{E_g} \tag{A.4.20}$$

where $U_a \cos\varphi = E_g - I_a R_a$. The output terminal power, P_{out}, and the back-EMF (rms value) of the no-load condition is calculated as eq.(A.4.21) and (A.4.22).

$$P_{out} = mI_a U_a \cos\varphi \tag{A.4.21}$$

$$E_0 = X_{dm} I_d + \sqrt{E_g^2 - \left(X_{qm} I_q\right)^2} = X_{dm} I_a \sin\psi + \sqrt{E_g^2 - \left(X_{qm} I_a \cos\psi\right)^2} \tag{A.4.22}$$

The flux density in the stator iron teeth is calculated as

$$B_t = \frac{1}{2}\left(\widehat{B}_g + \widehat{B}_{g,r}\right)\frac{\tau_s}{b_t} \tag{A.4.23}$$

where B_t is the flux density in the stator iron teeth, τ_s is the slot pitch, b_t is the teeth width. The average value of $\widehat{B}_g$ and $\widehat{B}_{g,r}$ is taken because the amplitude of rectangular flux waveform would underestimate the flux density in the iron teeth, while the amplitude of the sinusoidal flux waveform would overestimate the flux density in the iron teeth.

Another way to calculate the flux density in the stator iron teeth is as follow.

$$\varphi_t = \tau_s \widehat{B}_g l_s - (\tau_s - b_t) l_s B_u = B_t b_t l_s \tag{A.4.24}$$

where φ_t is the flux into the stator iron tooth, and B_u is the flux density in the slot area

Hence,

$$B_t = \frac{\tau_s}{b_t}\widehat{B}_g - \frac{\tau_s - b_t}{b_t} B_u \tag{A.4.25}$$

Since the tangential component of the field strength is continuous at the interface of the iron and the air, that is $H_t = H_u$,

Then, the B_t equation (A.4.25) can be rewritten as

$$B_t = \frac{\tau_s}{b_t}\widehat{B}_g - \frac{\tau_s - b_t}{b_t} \mu_0 H_t \tag{A.4.26}$$

The crossing point of this B_t in eq. (A.4.26) with B-H curve is the operating B_t, H_t, and μ_t. The flux of one pole, ϕ, in the air gap and the flux density in the stator yoke, B_{sy}, are

$$\phi = B_g \frac{\pi D_\delta l_s}{2p} = \left(\frac{2}{\pi}\hat{B}_g\right)\frac{\pi D_\delta l_s}{2p} = \left(\frac{2}{\pi}\hat{B}_g\right)\tau_p l_s \tag{A.4.27}$$

$$B_{sy} = \frac{\phi}{2h_{sy}l_s} = \frac{\hat{B}_g \tau_p}{\pi h_{sy}} \tag{A.4.28}$$

where, h_{sy} is the height of the stator yoke

The flux leakage of the rotor pole, ϕ_{pu}, is calculated as [33].

$$\phi_{pu} = 4\mu_0 l_s N_f I_f \left[\frac{h_{pc}}{3(\tau_p - b_{pc})} + \frac{h_{ps}}{\tau_p - b_{ps}}\right] \tag{A.4.29}$$

From the phasor diagram Fig. A. 4.2 it can be seen that the main flux in the air gap generated by the pole ϕ can be calculated by flux in the air gap at no-load condition ϕ_f.

$$\phi_f = \frac{\phi}{\cos\psi} \tag{A.4.30}$$

The total flux in the pole core ϕ_{pc} is the sum of the pole leakage flux ϕ_{pu} and flux in the air gap at no-load condition ϕ_f, which is

$$\phi_{pc} = \phi_f + \phi_{pu} = \frac{\left(\frac{2}{\pi}\hat{B}_g\right)\tau_p l_s}{\cos\psi} + 4\mu_0 l_s N_f I_f \left[\frac{h_{pc}}{3(\tau_p - b_{pc})} + \frac{h_{ps}}{\tau_p - b_{ps}}\right] \tag{A.4.31}$$

The flux density in the pole body, B_{pc}, is calculated as (A.4.32) and the flux density in the rotor yoke, B_{ry}, is calculated as (A.4.33), where h_{ry} is the height of the rotor yoke.

$$B_{pc} = \frac{\phi_{pc}}{l_s b_{pc}} = \frac{1}{b_{pc}}\left\{\frac{\left(\frac{2}{\pi}\hat{B}_g\right)\tau_p}{\cos\psi} + 4\mu_0 N_f I_f \left[\frac{h_{pc}}{3(\tau_p - b_{pc})} + \frac{h_{ps}}{\tau_p - b_{ps}}\right]\right\} \tag{A.4.32}$$

$$B_{ry} = \frac{B_{pc} b_{pc}}{2h_{ry}} \tag{A.4.33}$$

By using the flux density and *B-H* curve, the relative permeability of rotor core, rotor pole, stator teeth, and stator core, μ_{ry}, μ_{pc}, μ_t, μ_{sy}, and magnetic strength of each component H_{ry}, H_{pc}, H_t, H_{sy} can be calculated. Hence, the total ampere-turns (magnetomotive force) is

$$F_{total} = H_{ry}h_{ry} + H_{pc}\left(h_{pc} + h_{ps}\right) + H_g\delta_{eff} + H_t b_t + H_{sy}h_{sy} = \cos\psi N_f I_f^{'} \quad \text{(A.4.34)}$$

By comparing I_f and $I_f^{'}$, δ_{eff} is iterated to take the saturation into consideration. For example, if $I_f^{'} > I_f$, which means the setting MMF is not enough for the machine as iron cores are highly saturated, and the effective airgap is larger than the value used before. H_g is air gap magnetic strength which is calculated according to the rectangular flux density.

List of Figures

List of Tables

Abbreviations

AHP	Analytic hierarchy process
AMSC	American Superconductor
BSCCO	Barium strontium calcium oxocuprate
DC	Direct current
DFIG	Doubly fed induction generator
DP	Double pancake
EESG	Electrical excited synchronous generator
back-EMF	The back electromotive force
FEA	Finite element analysis
FPGA	Field programmable gate array
HTS	High temperature superconductor
LCOE	Levelized Cost of Electricity
MgB_2	Magnesium diboride
MMF	Magnetomotive force
MRI	Magnetic resonance imaging
PMSG	Permanent magnet synchronous generator
PSO	Particle swarm optimisation
ReBCO	Rare earth barium oxocuprates
SCoE	Society's Cost of Electricity

Index of symbols

A	Input power to the toroidal core (W)
A_a	Electrical loading of armature winding (A/m)
A_l	Cross section of the current leads (m^2)
A_{nm}	Coefficient to express φ_z due to the nth harmonic in region m
A_{rod}	Cross section per rod (m^2)
a	Number of branch pairs in armature winding
B	A term represents ohmic losses in the primary winding,
$B_0(T)$	Coefficients with temperature T in lift factor caculation
$\widehat{B}$	Magnitude of fundamental radial flux density (T)
B_δ	Average radial flux density in the middle of air-gap (T)
$B_{\delta max}$	Magnitude of the radial flux density in the middle of air gap (T)
$B_{\delta n}(x)$	Airgap radial flux of n-conductor in x position (T)
B_a	Applied field to the superconducting coils (T)
B_{an}	Applied field to the superconducting coils ($H//ab$) (T)
B_{ap}	Applied field to the superconducting coils ($H//c$) (T)
$B_{c0}(T)$	Coefficients with temperature T in lift factor caculation
B_m	Flux density in region m (T)
B_{mp}	Maximum flux density in field coil ($H//ab$) (T)
B_{mp1}	Maximum flux density in field coil (H//ab) due to a given current (T)
B_{mp2}	Crossing point of load line and 60% $I_c(B_a, T_{sc})$ line ($H//ab$) (T)
B_{mn}	Maximum flux density in superconducting field coil ($H//c$) (T)
B_{mn1}	Maximum flux density in field coil ($H//c$) due to a given current (T)
B_{mn2}	Crossing point of load line and 60% $I_c(B_a, T_{sc})$ line ($H//c$) (T)
B_r	Radial component of flux density (T)
B_{rm}	Radial component of flux density of region m (T)
B_{rmj}	Radial component of flux density in region m due to slot pair (j/j') (T)
B_{ry}	Magnetic flux density in the rotor yoke (T)
$B_{sc}(T)$	Coefficient related to the irreversibility field in calculation of $I_{cs}(B,T)$
B_{sy}	Magnetic flux density in the stator yoke (T)
B_t	Tangential component of flux density (T)
B_{tm}	Tangential component of flux density of region m (T)
B_{tmj}	Tangential component of flux density in region m due to slot (j/j') (T)
B_x	Parallel field component applied to the superconducting tapes (T)
B_y	Perpendicular field component applied to the superconducting tapes (T)
$b(T)$	Coefficients with temperature T in lift factor caculation
C	A term represents core magnetization and losses

C_a	Temperature coefficient of resistance (0.00393/°C)
C_{ec}	Coefficient of eddy current loss
C_{exc}	Coefficient of excess loss
C_{hyst}	Coefficient of hysteresis loss
C_{nm}	Coefficient to express φ_z due to the nth harmonic in region m
C_p	Rotor power coefficient
D_a	Outer diameter of the rotor (m), also called as armature diameter
D_r	Rotor diameter (m)
d_{ec}	Copper mass density (kg/m^3)
d_{esc}	Mass density of SC wires (kg/m^3)
d_y	Mass density of silicon lamination (kg/m^3)
E	Coefficient in flux density calculation
E_c	Corresponds to the critical electric field , which is 10^{-4} V/m
E_{steel}	Young's modulus for steel (Pa)
f	Frequency of the generator (Hz)
$f(\theta)$	Elliptical angular dependence equation of superconductor
f_e	Normalized efficiency
f_{rd}	Outward radial pressure (N/m^2)
f_{tl}	Normalized ratio of output torque to the length of tapes
f_{tm}	Normalized ratio of output torque to mass of active materials
f_{tv}	Normalized ratio of output torque to the volume
$f_{objective}$	Objective function to be optimized
f_{rdisc}	Uniform outward radial force per unit length of rotor disc (N/m)
f_{sdisc}	Uniform inward radial force per unit length of stator disc (N/m)
f_{sd}	Inward radial pressure (N/m^2)
$g_{gravity}$	Gravity acceleration (N/kg)
H_{tm}	Tangential component of magnetic strength of region m (A/m)
h_{sc}	Height of one side of superconducting field coil (m)
h_{scd}	Height of superconducting coil to stator core (m)
h_{sl}	Height of slot (m)
h_{ry}	Height of rotor core (m)
h_{sy}	Height of stator core (m)
I_{aN}	Rated current of armature winding (A)
I_c	Critical current of the superconducting tapes (A)
$I_c(B,T)$	Critical current as a function of temperature T and applied field B
$I_{cs}(B,T)$	Critical current due to strong linked current path (A)
$I_{cw}(B,T)$	Critical current due to weak linked current path (A)
$i_p(t)$	Primary winding current (A)
I_{scp}	Critical current due to maximum flux density ($H//ab$) (A)
I_{scn}	Critical current due to maximum flux density ($H//c$) (A)

I_{sc}	Current of the superconducting field coils (A)
I_{slot}	Current in a slot (A)
i_a	Current per turn in armature winding (A)
J_{aj}	Current density distribution of the armature in *j*/*j'* slot pair (A/m^2)
J_c	Critical current density with applied field *B* and temperature *T* (A/m^2)
J_c(77K, sf)	Critical current density at 77 K and only self field is considered (A/m^2)
J_{scz}	Circumferential current density of the superconducting winding (A/m^2)
j/*j'*	Slot pair number in the rotor
j_a	Current density in armature winding (A/m^2)
$j_{c0}(T)$	Coefficients with temperature *T* in lift factor caculation
k	Thermal conductivity of the titanium alloy (W/(K·m))
k_{cu}	Thermal conductivity of the copper (W/(K·m))
k_e	Coefficient to express φ_z due to the *n*th harmonic in region III
k_{eaj}	Coefficient to calculate flux density generated by slot pair *j*/*j'*
k_{Fe}	Space factor of the core
k_{m1}	An experimental factor in calculation of the sum of bearing friction losses, windage losses, and ventilator losses
k_{m2}	An friction coeficient for brush friction loss calculation
k_{ry}	Correction coefficient for iron losses calculation in rotor iron core
k_{sy}	Correction coefficient for iron losses calculation in stator iron core
$k(T)$	Coefficients with temperature *T* in lift factor caculation
L_0	Lorenz number, 2.45×10^{-8} WΩ/K^2
$L_0(T)$	Coefficients with temperature *T* in lift factor caculation
$L(B,T)$	Lift factor with applied magnetic field *B* and temperature *T*
L_{rod}	Length of the support rod (m)
L_{sc}	Length of superconducting tapes (m)
L_{scw}	Length of the superconducting coil per turn (m)
l_{aw}	Length of armature winding per coil (m)
l_{end}	Axial length of the end armature winding per turn in one side (m)
l_{endsc}	Axial length of one end of superconducting coil (m)
l_{ext}	Extension length of armature winding per coil in direction part (m)
l_{extsc}	Extension length of a SC coil in direction part (m)
l_r	Rotor length (m)
m	Number of region
m_{Cu}	Active material weight of copper winding (kg)
m_g	Active material weight of generator (kg)
m_{ry}	Active material weight of rotor core (kg)
m_{sc}	Active material weight of superconducting winding (kg)
m_{sy}	Active material weight of stator core (kg)
m_{frame}	structural material weight of the generator frame (kg)

n	Harmonic order
n-value	Indicating the sharpness of the *E-J* characteristic
N_a	Number of turns per armature coil
N_{cond}	Number of conductors in series
N_{sc}	Number of turns per superconducting coil
N_s	Number of turns of secondary winding
N_p	Number of turns of primary winding
n_N	Rated rotation speed (1/min)
P	Mechanical power (W)
P_{15}	Losses of ferromagnetic materials per mass at 1.5 T, 50 Hz (W/kg)
P_{add}	Additional losses (W)
P_b	Brushes losses (W)
P_{cu}	Copper losses (W)
P_{Fe}	Iron losses (W)
P_{fw}	Friction and windage loss (W)
P_{m1}	Sum of the bearing friction, windage, and ventilator losses (W)
P_{m2}	Brush friction losses (W)
P_N	Output power (W)
P_{rB}	Brush average pressure density (N/m^2)
P_{ac1}	Ramping current loss (W)
P_{ac2}	Ripple current loss (W)
P_c	Conduction heat leakage of current leads (W)
P_{cl1}	Heat load of current leads between 30 K and 80 K (W)
P_{cl2}	Heat load of current leads between 300 K and 80 K (W)
P_J	Ohmic losses of current leads (W)
P_m	Heat load due to measurement wires (W)
P_{r1}	Radiative load from the Dewar to the thermal shield (W)
P_{r2}	Radiative load to the coil (W)
P_{rod}	Heat leakage of support rods (W)
P_{rod1}	Heat leakage of support rods between 80 K and 300 K (W)
P_{rod2}	Heat leakage of support rods between 80 K and 30 K (W)
P_{s1}	Total heat load to 80 K (W)
P_{s2}	Total heat load to 30 K (W)
p	Number of pole pairs
p_g	Electromagnetic power(W)
q	Pole pair number
$\bar{q}$	Normal stress (N/m^2)
$q_{Nstress}$	Uniform normal stress (N/m^2)
R	Radius of the rotor balde (m)
R_a	Resistance of armature winding (Ω)

R_{at}	Resistance of an armature coil per turn (Ω)
R_{p}	Resistance of the primary winding (Ω)
R_{ry}	Radius of the rotor structure (m)
R_{sy}	Radius of the stator structure (m)
r	Radius of the calculated point (m)
r_{sci}	Inner radius of the superconducting coils (m)
r_{sco}	Outer radius of the superconducting coils
r_{shaft}	Radius of the rotor's shaft (m)
r_{ri}	Inner radius of the rotor core (m)
r_{ro}	Outer radius of the rotor core (m)
r_{si}	Inner radius of the stator core (m)
r_{sli}	Inner radius of the slots (m)
r_{slo}	Outer radius of the slots (m)
r_{so}	Outer radius of the stator core (m)
S	Number of slot pair per pole pair in the rotor
S_{sc}	Cross section area of SC tape (m^2)
t	An integer
T	Rated torque (Nm)
T^*	parameter in calculation of $I_c(B,T)$
T_1	Temperature of one of the rod or current lead (K)
T_2	Temperature of one of the rod or current lead (K)
T_a	Working temperature of armature winding (°C)
T_c	Critical temperature of the superconductor (K)
T_{da0}	Winding turn density in each slot (turn/m^2)
$T_{daj}(p\theta)$	Circumferential distribution of the winding turn density of slot pair *j/j'*
T_{dsc0}	Winding turn density of the superconducting coil (turn/m^2)
$T_{dsc}(p\theta)$	Circumferential distribution of the winding turn density of the superconducting coils per pole pair
T_{high}	High temperature side (K)
T_{low}	Low temperature side (K)
T_{sc}	Temperature of superconducting coils (K); A parameter to calculate $I_{cs}(0,T)$
t_{rdisc}	Thickness of the rotor disc (m)
t_{sdisc}	Thickness of the stator disc (m)
U_b	Voltage of brushes (V)
U_N	Rated voltage (V)
u	Number of coils per slot
u_{adflr}	Axial deflection of the rotor due to the gravity (m)
u_{rdflr}	Maximum radial deflection at both ends of the rotor (m)
u_{rdfls}	Radial deflection of the stator structure due to the normal stress

u_{adfls}	Axial deflection of the stator disc due to the gravity (m)
$u_{\mathrm{amp}}(t)$	Amplifier voltage (V)
$u_{\mathrm{s,ind}}(t)$	Induced voltage of the primary winding (V)
V_{aN}	Back-EMF of armature winding (V)
v	Velocity the conductor cuts through the magnetic field (m/s)
v_r	Surface speed of the rotor (m/s)
v_{rB}	Surface speed of the brush (m/s)
v_w	Wind velocity (m/s)
W_{ry}	Line load due to the weight of the rotor back yoke
W_{sy}	line load due to the weight of the stator back yoke
w_e	Weighting factor of normalized efficiency
w_{rdisc}	Uniformly distributed pressure due to the weight of the rotor disc
w_{sc}	Width of one side of superconducting field coil (m)
w_{sdisc}	Uniformly distributed pressure due to the weight of the stator disc
w_{sl}	Width of slot (m)
w_{tl}	Weighting factor of normalized ratio of the torque to length of tapes
w_{tm}	Weighting factor of normalized ratio of the torque to active mass
w_{tv}	Weighting factor of normalized ratio of the torque to volume
X	Original value of the dataset
X'	Normalized value of the dataset
X_{max}	Maximum value of the dataset to be normalized
X_{min}	Minimum value of the dataset to be normalized
y_1	Coil span as the number of coil sides
Z_u	Number of slots in the generator
α	parameter in calculation of $I_c(B,T)$
α_1	Half angles between two adjacent coil sides from different coils (◦)
α_2	Half angles between two adjacent coil sides of one coil (◦)
α_{f1}	Safety margin for SC coils
α_{f2}	Packing factor for SC coils
α_{DC}	Relative pole width for DC machines
α_{sl}	Slot spreading angle (◦)
α_u	Slot filling factor
α_p	Pole arc coefficient
$\alpha(T)$	Coefficient related to thermally activated flux creep
β	Coefficients with junction geometry in lift factor caculation
β_{aend}	Bending angle of the armature end winding to the coil span (◦)
β_j	Mechanical angle of slot pair j/j' shifting from slot pair(1/1') (◦)
δ	Physical air-gap length (m)
δ_{em}	Effective air-gap length (m)
δ_{te}	Air gap that takes the iron teeth into account (m)

η	Assumed efficiency of the generator
η_c	Carnot efficiency
η_g	Calculated generator efficiency
θ	Mechanical angle (∘) In terms of applied field to tape, the origin of the angle θ is for an applied field parallel to the *ab* planes (∘)
κ	Coefficient in Carter factor calculation
κ_c	Carter factor
μ_{rr}	Relative permeability of rotor core
μ_{rr0}	Initial relative permeability of rotor core
μ_{rr1}	Calculated relative permeability of rotor core
μ_{rs}	Relative permeability of stator core
μ_{rs0}	Initial relative permeability of stator core
μ_{rs1}	Calculated relative permeability of stator core
μ_0	Permeability of air (Vs/Am)
ρ_{20}	Resistivity of copper in 20 Celsius, 1.724E-08 Ωm
ρ	Resistivity of the material
ρ_{air}	Air density (kg/m^3)
ρ_{cu}	Electrical resistivity of the copper (Ωm)
ρ_{sc}	Resistivity of the superconductor (Ωm)
ρ_{steel}	Mass density of the steel (kg/m^3)
τ_p	pole pitch (m)
τ_u	Slot pitch (m)
τ_y	Distance of the coil span (m)
φ	Tilted angle of the axis of the rotor blades and generator (∘)
φ_m	Flux linkage per pole (Wb)
φ_z	Magnetic vector potential in *z*-direction (Vm/s)
φ_{zm}	Magnetic vector potential of region *m* (Vm/s)
Ω_N	Rated mechanical angular speed (rad/s)
$\overline{\sigma}$	Shear stress (N/m^2)
v_{steel}	Poisson's ratio
λ	Tip speed ratio
λ_{ti}	Thermal conductivity of titanium alloy (W/(K·m))
θ_p	Pitch angle
γ	Parameter in calculation of $I_{cs}(0,T)$

Publications

[1] **Yingzhen Liu**, Jing Ou, Markus Schiefer, Patrick Breining, Francesco Grilli and Martin Doppelbauer, "Application of an Amorphous Core to an Ultra-high-speed Sleeve-free Interior Permanent-magnet Rotor," *Trans. on Ind. Electron*, vol. 28. no. 3, 2018.

[2] **Yingzhen Liu**, Mathias. Noe, Jing Ou, Patrick Breining, Marc Veigel, and Martin Doppelbauer, "Measurement of magnetic materials at room and cryogenic temperature for their application on superconducting wind generators," *IEEE Trans. Appl. Supercond.*, vol. 28. no. 3, 2018, Art. ID. 5206006.

[3] **Yingzhen Liu**, Jing Ou, and Mathias Noe, "A large-scale superconducting DC wind generator considering concentrated/distributed armature winding," *IEEE Trans. Appl. Supercond.*, vol. 27, no.4, Jun. 2017, Art. ID. 5200405.

[4] **Yingzhen Liu**, Mathias Noe, and Martin Doppelbauer, "Feasibility study of a superconducting DC direct-drive wind generator," *IEEE Trans. Appl. Supercond.*, vol. 26, no.4, Jun. 2016, Art. ID. 5203906.

[5] **Yingzhen Liu**, Jing Ou, and Mathias Noe, "Integrated electromagnetic-mechanical design method for optimized superconducting DC Wind Generators," presented at 15th Wind International Workshop, Vienna, Nov. 2016

[6] **Yingzhen Liu**, Ronghai Qu, Jin Wang, Haiyang Fang, Xiaolong Zhang, and Hong chen, "Influences of generator parameters on fault current and torque in a large-scale superconducting wind generator," *IEEE Trans. Appl. Supercond.*, vol. 25, no.6, Dec. 2015, Art. ID. 5204309.

[7] **Yingzhen Liu**, and Mathias Noe, "Analytical method to calculate flux density in a superconducting (SC) direct-current (DC) wind generator," presented at 2015 IEEE Magnetics Conference (INTERMAG), Bejing, May 2015.

[8] Jing Ou, **Yingzhen Liu**, Ronghai Qu and Martin Doppelbauer, "Experimental and theoretical research on cogging torque of PM synchronous motors considering manufacturing tolerances," *Trans. on Ind. Electron*, vol. 65, no. 5, pp. 3772-3783, 2018.

[9] Jing Ou, **Yingzhen Liu**, Markus Schiefer, and Martin Doppelbauer, "A novel PM-free high-speed linear machine with amorphous primary core," *IEEE Trans. on Magn.*, vol. 53. no. 11, 2017, Art. ID. 4002108.

[10] Patrick Breining, Marc Veigel, Martin Doppelbauer, **Yingzhen Liu** and M. Noe, "Iron loss measurement of nonoriented silicon and cobalt iron electrical steel sheets at liquid nitrogen temperature using ring specimen," presented at IEEE International Electric Machines and Drives Conference (IEMDC), Miami, FL, USA, 2017.

[11] Jin Wang, Ronghai Qu, Yuejin Tang, **Yingzhen Liu**, Bin Zhang, Jie He, Zhe Zhu, Haiyang Fang, and Lushun Su, "Design of a superconducting synchronous generator

with LTS field windings for 12 mw offshore direct-drive wind turbines," *IEEE Trans. on Ind. Electron.*, vol. 63, no. 3, pp.1618-1628, Mar. 2015

Bibliography

[1] Global Wind Energy Council, "Global wind report," Technical report, 2015. [Online]. Available: http://www.gwec.net/wp-content/uploads/vip/GWEC-Global-Wind-2015-Report_April-2016_19_04.pdf

[2] Global Wind Energy Council, "Global wind energy outlook," Technical report, 2015. [Online]. Available: http://www.gwec.net/publications/global-wind-energy-outlook/global-wind-energy-outlook-2016/

[3] B. Maples, M. Hand, and W. Musial, Comparative assessment of direct drive high temperature superconducting generators in multimegawatt class wind turbines, Nat. Renew. Energy Lab., Golden, CO, Rep. NREL/TP-5000-49086, 2010. [Online]. Available: http://www.osti.gov/bridge/ product.biblio.jsp?osti_id=991560

[4] G. Snitchler. "Progress on High Temperature Superconductor Propulsion Motors and Direct Drive Wind Generators," IPEC, Sapporo, 2010.

[5] M. Liserre, R. Cardenas, M. Molinas, and J. Rodriguez, "Overview of multi-MW wind turbines and wind parks," *IEEE Trans. Ind. Electron.*, vol. 58, no. 4, pp. 1081-1095, Apr. 2011.

[6] J. Robinson, D. Jovcic and G. Joos, "Analysis and Design of an Offshore Wind Farm Using a MV DC Grid," *IEEE Trans. Power Delivery*, vol. 25, no. 4, pp. 2164-2173, Oct. 2010.

[7] C. Meyer, M. Hoing, A. Peterson and R. W. De Doncker, "Control and Design of DC Grids for Offshore Wind Farms," *IEEE Trans. Ind. Appl.*, vol. 43, no. 6, pp. 1475-1482, Nov./Dec. 2007.

[8] C. Chen, A. Q. Huang, C. Li, G. Wang and W. Gu, "Analysis and Comparison of Medium Voltage High Power DC/DC Converters for Offshore Wind Energy Systems," *IEEE Trans. Power Electron.*, vol. 28, no. 4, pp. 2014-2023, April 2013.

[9] E. Spooner, and B. J. Chamers, "DC motors with high-critical-temperature superconducting field winding and slotless armature," *IEE Proc.-Electr. Power Appl.*, vol. 141, no. 3, pp. 135-143, May, 1994.

[10] David Milborrow, "Wind economics," *Wind Power Monthly*, vol. 30, no. 6, pp. 41, Jun. 2016.

[11] Global Wind Energy Council, "Global wind energy outlook," Technical report, 2014. [Online]. Available: http://www.gwec.net/wpcontent/uploads/2015/03/GWEC_Global_Wind_2014_Report_LR.pdf

[12] Comment: Offshore wind's new benchmark, *Wind Power Monthly*, vol. 30, no. 6, pp. 4, Aug. 2016.

[13] H. Polinder *et al.*, "Trends in wind turbine generator systems," *IEEE J. Emerging Sel. Topics Power Electron.*, vol. 1, no. 3, pp. 174-185, Sep. 2013.

[14] J. Wang *et al.*, “Design of a superconducting synchronous generator with LTS field windings for 12 MW offshore direct-drive wind turbines”, *IEEE Trans. Ind. Electron.*, vol. 63, no. 3, pp. 1618-1628, 2016.

[15] The European Wind Energy Association, “The European offshore wind industry - key trends and statistics 2015,” Technical report, 2016. [Online]. Available: http://www.ewea.org/fileadmin/files/library/publications/statistics/EWEA-European-Offshore-Statistics-2015.pdf

[16] MHI Vestas, *V164 8.0MW* [Online]. Available. http://www.homepages.ucl.ac.uk/~uceseug/Fluids2/Wind_Turbines/Turbines/V164-8MW.pdf

[17] Shaun Campbell, “Ten of the biggest wind turbines,” *Windpower Monthly*, vol. 32, no. 7, pp. 24-28, Jul. 2016.

[18] 4C Offshore, *V164-8.0 MW* [Online]. Available. http://www.4coffshore.com/windfarms/turbine-mhi-vestas-offshore-wind-v164-8.0-mw-tid89.html

[19] Adwen, *8 MW Platform* [Online]. Available. http://www.adwenoffshore.com/products-services/products/8-mw-turbines/

[20] 4C Offshore, *AD8-180* [Online]. Available. http://www.4coffshore.com/windfarms/turbine-adwen-ad-8-180-tid197.html

[21] Enercon. *E 126* [Online]. Available. http://www.enercon.de/en/products/ep-8/e-126/

[22] Wind turbine model.com, *Siemens SWT-7.0-154* [Online]. Available. http://en.wind-turbine-models.com/turbines/1102-siemens-swt-7.0-154

[23] 4C Offshore, *SWT-7.0-154* [Online]. Available. http://www.4coffshore.com/windfarms/turbine-siemens-swt-7.0-154-tid205.html

[24] Windpower Monthly, *Ming Yang completes 6.5MW offshore turbine* [Online]. Available.http://www.windpowermonthly.com/article/1188373/ming-yang-complet-es-65mw-offshore-turbine

[25] 4C Offshore, *SCD 6MW* [Online]. Available. http://www.4coffshore.com/windfarms/turbine-aerodyn-engineering-gmbh-scd-6mw-tid202.html

[26] Senvion, *6.2M152 - the next generation of offshore turbines* [Online]. Available. https://www.senvion.com/global/en/wind-energy-solutions/wind-turbines/6xm/62m152/

[27] 4C Offshore, *6.2M125* [Online]. Available. http://www.4coffshore.com/windfarms/turbine-senvion-6.2m-152-tid192.html

[28] GE Reports, *What America's First Offshore Wind Farm Reveals About GE's Alstom Deal* [Online]. Available. http://www.gereports.com/wind-farm-off-rhode-islands-coast-reveals-ges-alstom-acquisition/

[29] 4C Offshore, *SL6000/155* [Online]. Available. http://www.4coffshore.com/windfarms/turbine-sinovel-sl6000-155-tid159.html

[30] 4C Offshore, *UP6000-136* [Online]. Available. http://www.4coffshore.com/windfarms/turbine-guodian-united-power-up6000-136-tid128.html

[31] Siemens, Wind Turbine SWT-6.0-154 Technical specifications, [Online]. Available. https://www.siemens.com/content/dam/internet/siemens-com/global/market-specific-solutions/wind/data_sheets/data-sheet-wind-turbine-swt-6.0-154.pdf

[32] Enercon. *WEC components-Constant further development* [Online]. Available. http:// www.enercon.de/en/technology/wec-components/

[33] J. Pyrhonen, T. Jokinen, V. Hrabovcova, *Design of rotating electrical machines*, 2nd ed. West Sussex , United Kingdom: John Wiley & Sons Ltd, 2014.

[34] A. Petersson, "Analysis, modeling and control of doubly-fed induction generators for wind turbines," Ph.D. dissertation, Chalmers Univ., Göteborg, Sweden,2005.

[35] G. Michalke, "Variable speed wind turbines-Modelling, control, and impact on power systems," Ph.D. dissertation, TU Darmstadt, Germany, 2008.

[36] H. Li and Z. Chen, "Overview of different wind generator systems and their comparisons," *IET Renew. Power Gener.*, vol. 2, no. 2, pp. 123-138, Jun. 2008.

[37] D. Bang, "Design of transverse flux permanent magnet machines for large direct-drive wind turbines," Ph.D. dissertation, Electr. Eng. Dept., Delft Univ. Technol., Delft, The Netherlands, 2010.

[38] M. Mueller and A. Zavvos, "Electrical generators for direct drive systems: A technology overview," in *Electrical Drives for Direct Drive Renewable Energy Systems*, Oxford, U.K.: Woodhead, 2013.

[39] H. Polinder, F. F. A. van der Pijl, G. J. de Vilder, and P. Tavner, "Comparison of direct-drive and geared generator concepts for wind turbines," *IEEE Trans. Energy Convers.*, vol. 21, no. 3, pp. 725-733, Sep. 2006.

[40] H. Polinder, "Overview of and trends in wind turbine generator systems," presented at 2011 IEEE Power Eng. Soc. General Meeting, San Diego, CA, USA, Jul. 2011.

[41] H. S. Kim and D. D. C. Lu, "Review on wind turbine generators and power electronic converters with the grid-connection issues," presented at 20th Universities Power Engineering Conference (AUPEC),Australasian, 2010.

[42] X. Yang, D. Patterson, and J. Hudgins, ''Permanent magnet generator design and control for large wind turbines,'' presented at IEEE Symp. Power Electron. Mach. Wind Appl. (PEMWA), Denver, CO, USA, Jul. 2012.

[43] V. Yaramasu, B. Wu, P. C. Sen, S. Kouro, and M. Narimani, "High power wind energy conversion systems: State-of-the-art and emerging technologies," *Proc. IEEE*, vol. 103, no. 5, pp. 740-788, May 2015.

[44] Narayan Poudyal and J Ping Liu, "Advances in nanostructured permanent magnets research," *J. Phys. D: Appl. Phys.*, vol. 46, no. 4, 2013, Art. ID. 043001.

[45] W. T. Benecki, "The permanent magnet market-2015," presented at Magnetics 2013 Conference, Orlando, FL, USA, 2013. [Online]. Available: http://www.waltbenecki.com/uploads/Magnetics_2013_Benecki_Presentation.pdf

[46] E-Mangets UK, *Grades of ferrite* [Online]. Available: http://www.ferrite-info.com/ferrite_grades.aspx

[47] ChenYang Technology, *SmCo Magnets-samarium kobalt magnets* [Online]. Available: http://www.smcomagnets.de/DE_Index.htm

[48] Barbara Slusarek and Kazimierz Zakrzewski, "Magnetic properties of permanent magnets for magnetic sensors working in wide range of temperature," *Przegląd Elektrotechniczny (Electrical Review)*,vol. 88 NR, no.7b, pp.123-126, 2012.

[49] A. M. El-Refaie, "Fractional-slot concentrated-windings synchronous permanent magnet machines: Opportunities and challenges," *IEEE Trans. Ind. Electron.*, vol. 57, no. 1, pp. 107-121, Jan. 2010.

[50] J.Cros, P.Viarouge, Y.Chalifour, and J.Figueroa, "A new structure of universal motor using soft magnetic composites," *IEEE Trans. Ind. Appl.*, vol. 40, no. 2, pp. 550-557, 2004.

[51] J. Wang, R. Qu, and Y. Liu, "Study of multiphase superconducting wind generators with fractional-slot concentrated windings," *IEEE Trans. Appl. Supercond.*, vol. 24, no. 3, Jun. 2014, Art. ID. 5202106.

[52] R. Dutta, L. Chong, and M. F. Rahman, "Design and experimental verification of an 18-slot/14-pole fractional-slot concentrated winding interior permanent magnet machine," *IEEE Trans. Energy Convers.*, vol. 28, no. 1, pp. 181-190, Mar. 2013.

[53] J. Yang, G. Liu, W. Zhao, Q. Chen, Y. Jiang, L. Sun, et al., "Quantitative comparison for fractional-slot concentrated-winding configurations of permanent-magnet vernier machines," *IEEE Trans. Magn.*, vol. 49, no. 7, pp. 3826-3829, Jul. 2013.

[54] D. Bang, H. Polinder, G. Shrestha, and J. A. Ferreira, "Review of generator systems for direct-drive wind turbines," presented at Eur. Wind Energy Conf. Exhib., Brussels, Belgium, 2008.

[55] M. Dubois, "Optimized permanent magnet generator topologies for direct drive wind turbines", Ph.D. dissertation, Delft University of Technology, Delft, The Netherlands, 2004.

[56] D. Bang, H. Polinder, G. Shrestha, and J.A. Ferreira, "Promising direct drive generator system for large wind turbines," Presented at 2008 Wind Power to the Grid - EPE Wind Energy Chapter 1st Seminar, Delft, The Netherlands, 2008.

[57] A.S. McDonald, M.A. Mueller and H. Polinder, "Comparison of generator topologies for direct-drive wind turbines including structural mass", Presented at The International Conference on Electrical Machines (ICEM), Chania, Crete Island, Greece, 2006.

[58] D. Bang, H. Polinder, G. Shrestha, and J.A. Ferreira, "Comparative design of radial and transverse flux PM generators for direct-drive wind turbines", presented at 2008 International Conference on Electrical Machines, Vilamoura, Portugal, Sep. 2008.

[59] EWEA, "Upwind: Design limits and solutions for very large wind turbines," Tech. rep., March 2011. [Online]. Available: http://www.ewea.org/fileadmin/files/library /publications/reports/UpWind_Report.pdf

[60] DTU Wind, "INNWIND.EU: Design and PI of PDD generator," Tech. rep., 2015, [Online]. Available: http://www.innwind.eu/-/media/Sites/innwind/Publications /D-eliverables/INNWIND-EU-Deliverable-D3-21-Final-report1.ashx?la=da

[61] R. Fair, "Superconductivity for large scale wind turbines," GE Global Res., Niskayuna, NY, USA, Tech. Rep. DE-EE0005143, Apr. 2012.

[62] J. Lloberas, A. Sumper, M. Sanmarti, and X. Granados, "A review of high temperature superconductors for offshore wind power synchronous generators," *Renew. Sustain. Energy Rev.*, vol. 38, pp. 404-414, Oct. 2014.

[63] G. Snitchler, B. Gamble, C. King, and P. Winn, "10 MW class superconductor wind turbine generators," *IEEE Trans. Appl. Supercond.*, vol. 21, no. 3, pp. 1089-1092, Jun. 2011.

[64] I. Marino, A. Pujana, G. Sarmiento, S. Sanz, J. M. Merino, M. Tropeano, J. Sun, and T. Canosa, "Lightweight MgB_2 superconducting 10 MW wind generator," *Supercond. Sci. Technol.*, vol. 29,no. 2, Feb. 2016, Art. ID. 024005.

[65] Community Research and Development Information Service (CORDIS), *EcoSwing-Energy Cost Optimization using Superconducting Wind Generators-World's First Demonstration of a 3.6 MW Low-Cost Lightweight DD Superconducting Generator on a Wind Turbine* [Online]. Available: http://cordis.europa.eu/ project/rcn/ 195203_en.html

[66] A. M. Campbell, "Superconducting and conventional machines," *Supercond. Sci. Technol.*, vol. 27, no. 12, Dec. 2014, Art. ID. 124012.

[67] S. S. Kalsi, *Applications of High Temperature Superconductors to Electric Power Equipment*, Singapore: IEEE Press, 2011

[68] Sumitomo Electric, *New DI-BSCCO series* [Online]. Available: http://global-sei.com/super/hts_e/index.html

[69] "Special Session: What will drive market maturity for HTS applications in the electric power sector?," presented at Applied Superconductivity Conference, Sep. 2016, Denver, USA.

[70] SuperOx, *2G HTS* [Online]. Available: http://www.superox.ru/en/products/974-vtspprovod/

[71] THEVA, *THEVA Pro-Line* [Online]. Available: http://theva.com/superconductors /theva-pro-line/

[72] Deutsche Nanoschicht, *High temperature superconductor tape for commercial applications* [Online]. Available: http://d-nano.com/en/activities/superconductors. html

[73] SuperPower, *2G HTS wire* [Online]. Available: http://www.superpower-inc.com /content/2g-hts-wire

[74] Bruker, *YBCO HTS* [Online]. Available: https://www.bruker.com/products/superconductors-and-metal-composite-materials/superconductors/ybco-2g-hts-superconductors/learn-more.html

[75] AMSC, *HTS wire* [Online]. Available: http://www.amsc.com/solutions-products/hts_wire.html

[76] STI, *CUNDUCTUS®* [Online]. Available: http://www.suptech.com/what_is_conductus_n.php

[77] Columbus superconductors, *Products* [Online]. Available: http://www.columbus-superconductors.com/products.asp

[78] J. Pitel, P. Ková˘c, M. Tropeano, and G. Grasso, "Study of the potential of three different MgB2 tapes for application in cylindrical coils operating at 20 K," *Supercond. Sci. Technol.*, vol. 28, no. 5, May 2015, Art. ID 055012.

[79] S.I. Schlachter, W. Goldacker, MgB2 and BSCCO, presented at HTS ^{4}FUSION Conductor Workshop, Karlsruhe, Germany, May 2011.

[80] Minwon Park, Realization of a large-scale superconducting generator for a wind power generation system, presented at ESAS Summer School on HTS Technology for Sustainable Energy and Transport System, Bologna, Italy, Jun. 2016.

[81] Y. Yamada, B. Marchionini, *High Temperature Superconductivity, A Roadmap for the Electric Power Sector, 2015-2030* [Online]. Available: http://www.ieahts.org/Publications.

[82] S. S. Kalsi, "Development status of superconducting rotating machines," presented at IEEE PES Meeting, New York, 2002.

[83] K. Ueda, T. Ageta and S. Nakayama, "Super-GM and other superconductivity projects in Japanese electric power sector," *IEEE Trans. Appl. Supercond.*, vol. 7, no. 2, pp. 245-251, June 1997.

[84] Wei Tong, *Wind power generation and wind turbine design*, Southampton: WIT Press, 2010.

[85] G. Snitchler, Progress on high temperature superconductor propulsion motors and direct drive wind generators, Presented at The International Power Electronics Conference - ECCE ASIA, Sapporo, 2010.

[86] Ecoswing, *Project* [Online]. Available: http://ecoswing.eu/project

[87] O. Keysan, and M. A. Mueller, "Superconducting generators for renewable energy applications," Presented at IET Conference on Renewable Power Generation (RPG 2011), Edinburgh ,UK, Sep. 2011.

[88] S. S. Kalsi, "Superconducting wind turbine generator employing MgB_2 winding both on rotor and stator," *IEEE Trans. Appl. Supercond.*, vol. 24, no. 1, 2014, Art. ID. 5201907.

[89] D. Liu, H. Polinder, N. Magnusson, J. Schellevis, and B. Abrahamsen, "Ripple field AC losses in 10-MW wind turbine generators with MgB2 superconducting field winding," *IEEE Trans. Appl. Supercond.*, vol. 26, no. 3, 2016, Art. ID. 5204205.

[90] ECO 5, "Kostengünstige Supraleitergeneratoren auf MgB2-Basis für Massenmärkte der Windenergie," ECO 5, Bonn, Germany, Tech. Rep. DBU AZ31934, 2016.

[91] R. Qu, Y. Liu and J. Wang, "Review of superconducting generator topologies of direct-drive wind turbines," *IEEE Trans. Appl. Supercond.*, vol. 23, no. 3, 2013, Art. ID. 5201108.

[92] R. Shafaie and M. Kalantar, "Design of a 10-MW-class wind turbine HTS synchronous generator with optimized field winding," *IEEE Trans. Appl. Supercond.*, vol. 23, no. 4, 2013, Art. ID. 5202307.

[93] Y. Liu, R. Qu and J. Wang, "Comparative analysis on superconducting direct-drive wind generators with iron teeth and air-gap winding," *IEEE Trans. Appl. Supercond.*, vol. 24, no. 3, 2014, Art. ID. 5201405.

[94] G. Klaus, M. Wilke, J. Frauenhofer, W. Nick and H. W. Neumuller, "Design challenges and benefits of HTS synchronous machines," presented at IEEE Power Eng. Soc. Gen. Meeting, Tampa, Jun. 2007.

[95] M. Hsieh, C. Lin and I. Lin, "Design and analysis of high temperature superconducting generator for offshore wind turbines," *IEEE Trans. Magn.* vol. 49, no. 5, pp. 1881-1884, 2013.

[96] Ozan Keysan, "Superconducting generators for large offshore wind turbines," Ph.D. dissertation, The University of Edinburgh, Edinburgh, UK, 2014.

[97] Y. Liu, M. Noe, and M. Doppelbauer, "Feasibility study of a superconducting DC direct-drive wind," *IEEE Trans. Appl. Supercond.*, vol. 26, no.4, Jun. 2016, Art. ID. 5203906.

[98] A. B. Abrahamsen, N. Magnusson, D. Liu, E. Stehouwer, H.B. Hendriks, and H. Polinder, "Design study of a 10 MW MgB2 superconductor direct drive wind turbine generator", in *Proceedings of the EWEA Conference*, Barcelona, 2014. [Online]. Available: http://orbit.dtu.dk/files/89914021/Design_study_of_a_10_MW .pdf

[99] A. B. Abrahamsen, N. Magnusson, B. B. Jensen, D. Liu, and H. Polinder, "Design of an MgB2 race track coil for a wind generator pole demonstration," *Journal of Physics: Conference Series*, vol. 507, no. 3, Sep. 2014, Art. ID. 032001.

[100] Sung H-J, Park M, Go B-S and Yu I-K, "A study on the required performance of a 2G HTS wire for HTS wind power generators," *Supercond. Sci. Technol.*, vol. 29, no. 5, 2016, Art. ID. 054001.

[101] H. Yamasaki and M. Furuse, "Feasibility study project to realize the merits of 10 MW class superconducting wind turbine generators," presented at the CEC/ICMC2015, Tucson, Arizona, U.S.A., Jul., 2015.

[102] C. Lewis and J. Müller, "A direct drive wind turbine HTS generator," Presented at IEEE Power Engineering Society General Meeting, Tampa, FL, 2007.

[103] Ozan Keysan and Markus Mueller, "A modular and cost-effective superconducting generator design for offshore wind turbines," *Supercond. Sci. Technol.*, vol. 28, 2015, Art. ID. 034004.

[104] Ainhoa Pujana, "Lightweight 10MW MgB2 superconducting generator", presented at Bilbao Marine Energy Week 2015, Bilbao, April 2015.

[105] H. Polinder, D. Bang, R.P.J.O.M. van Rooij, A.S. McDonald, and M.A. Mueller, "10 MW wind turbine direct-drive generator design with pitch or active speed stall control," Presented at IEEE Int. Electric Machines and Drives Conf., Antalya, Turkey, May 2007.

[106] Ozan Keysan and Markus A. Mueller, "A homopolar HTSG topology for large direct-drive wind turbines," *IEEE Trans. Appl. Supercond.*, vol. 21, no. 5, pp.3523-3531, 2011.

[107] S. Engström and S. Lindgren, "Design of NewGen direct-drive generator for demonstration in a 3.5 MW wind turbine", presented at EWEC (European Wind Energy Conference & Exhibition, Milan, Italy, May 2007.

[108] Z. Jiang, Y. Zhu and J. Jin, "Design process and performance analysis of a short axis 10 MW HTS wind generator," presented at IEEE International Conference on Applied Superconductivity and Electromagnetic Devices (ASEMD),Beijing, 2013.

[109] N. Maki, "Design study of high-temperature superconducting generators for wind power systems," *J. Phys. Conf. Ser.,* vol. 97, no. 1, Feb. 2008, Art. ID 012155.

[110] H-J Sung, M Park, I-K Yu, "Designs of 10 MW air-core and iron-core HTS wind power generators," *J. Electr. Eng. Technol.* vol. 10, no. 2, pp. 545-550, 2015.

[111] Versteegh C.J.A.,"Design of the Zephyros Z72 wind turbine with emphasis on the direct drive PM generator," presented at Nordic Workshop on Power and Industrial Electronics (NORPIE), Trondheim, Norway, June 2004.

[112] Satoshi Fukui, Jun Ogawa, Takao Sato, Osami Tsukamoto, Naoji Kashima, and Shigeo Nagaya, "Study of 10 MW-class wind turbine synchronous generators with HTS field windings," *IEEE Trans. Appl. Supercond.*, vol. 21, no. 5, pp.1151-1154, 2011.

[113] H. Ohsaki, Y. Terao, R. M. Quddes, and M. Sekino, "Electromagnetic characteristics of 10MW class superconducting wind turbine generators," Presented at 2010 International Conference on Electrical Machines and Systems, Incheon, 2010.

[114] Y. Xu, N. Maki, and M. Izumi, "Operating temperature influence on performance of 10 MW wind turbine HTS generators," *IEEE Trans. Appl. Supercond.*, vol. 25, no. 3, 2015, Art. ID 5200605.

[115] Sang-Ho Lee, Jung-Pyo Hong, Young-Kil Kwon, Young-Sik Jo, and Seung-Kyu Baik, "Study on homopolar superconductivity synchronous motors for ship propulsion applications," *IEEE Trans. Appl. Supercond.*, vol. 18, no. 2, pp. 717-720, 2008.

[116] Hae-Jin Sung, Gyeong-Hun Kim, Kwangmin Kim, Sung-Jun Jung, Minwon Park, In-Keun Yu, Young-Gyun Kim, Haigun Lee, and A.-Rong Kim, "Practical Design of a 10 MW superconducting wind power generator considering weight issue" *IEEE Trans. Appl. Supercond.*, vol. 23, no. 3, 2013, Art. ID 5201805.

[117] Y. Duan and R. G. Harley, "Present and future trends in wind turbine generator designs," Presented at 2009 IEEE Power Electronics and Machines in Wind Applications, Lincoln, NE, 2009.

[118] GE, *The temper of turbine: one of these wind turbines can power 5,000 homes* [Online]. Available: http://www.gereports.com/where-ge-makes-haliade-turbines/

[119] Enercon, *ENERCON product overview* [Online]. Available: http://www. enercon.de/fileadmin/Redakteur/Medien-Portal/broschueren/pdf/en/ENERCON_Produkt_en_06_2015.pdf

[120] INNWIND, *INNWIND Leaflet 2016* [Online]. Available. http://www.inn- wind.eu/

[121] The Crown Estate, "Offshore wind cost reduction pathways study," Technical report, 2012. [Online]. Available: ttps://www.thecrownestate.co.uk/media/5493/ei-offshore-wind-cost-reduction-pathways-study.pdf

[122] G. Bao, D. Zhang, J. Shi and J. Jiang, "Optimal design for cogging torque reduction of transverse flux permanent motor using particle swarm optimization algorithm," presented at The 4th International Power Electronics and Motion Control Conference, Xi'an, 2004.

[123] Mahdi Ashaani, and Yasser Abdel-Rady I. Mohamed, "Multiobjective shape optimisation of segmented pole permanent-magnet synchronous machines with improved torque characteristics," *IEEE Trans. Magn.*, vol. 47, no. 4, pp. 795-804, Apr. 2011.

[124] Hany M. Hasanien, "Particle swarm design optimisation of transverse flux linear motor for weight reduction and improvement of thrust force," *IEEE Trans. Ind. Electron.*, vol. 58, no. 9, pp. 4048-4056, Sep. 2011.

[125] S. Andrew Semidey, Yao Duan, J. Rhett Mayor, Ronald G. Harley, and Thomas G. Habetler, "Optimal electromagnetic-thermo-mechanical integrated design candidate search and selection for surface-mount permanent-magnet machines considering load profiles," *IEEE Trans. Ind. Appl.*, vol. 47, no. 6, pp. 2460-2468, 2011.

[126] A. A. Arkadan, M. N. ElBsat and M. A. Mneimneh, "Particle swarm design optimization of ALA rotor SynRM for traction Applications," *IEEE Trans. Magn.*, vol. 45, no. 3, pp. 956-959, March 2009.

[127] Y. Duan, R. G. Harley and T. G. Habetler, "Comparison of particle swarm optimization and genetic algorithm in the design of permanent magnet motors," presented at 2009 IEEE 6th International Power Electronics and Motion Control Conference, Wuhan, 2009.

[128] B. N. Cassimere and S. D. Sudhoff, "Population-based design of surface-mounted permanent-magnet synchronous machines," *IEEE Trans. Energy Convers.*, vol. 24, no. 2, pp. 338-346, June 2009

[129] Serkan Kiranyaz, Turker Ince, and Moncef Gabbouj, *Multidimensional particle swarm optimization for machine learning and pattern recognition*, New York, NY, USA: Springer, Aug. 2013.

[130] L. Liu, D. A. Cartes and W. Liu, "Particle Swarm Optimization Based Parameter Identification Applied to PMSM," presented at 2007 American Control Conference, New York, NY, 2007.

[131] A. Hughes and T. J. E. Miller, "Analysis of fields and inductances of air-cored and iron-cored synchronous machines," *Proc. IEE*, vol. 124, no. 2, pp. 121-126, Feb. 1977.

[132] S. K. Safi and J. R. Bumby, "Analysis of magnetic fields in the slotted structure of a superconducting AC generator," *Proc. IEE*, vol. 139, no. 5, pp. 411-422, Sep. 1992.

[133] M. Yazdanian, P. Elhaminia, M. R. Zolghadri,and M. Fardmanesh, "Analytical modeling of magnetic flux in superconducting synchronous machine," *IEEE Trans. Appl. Supercond.*, vol. 23, no. 1, Feb. 2013, Art. ID. 5200406.

[134] W. Zhang, D. Xia, D. Zhang, and G. Zhang, "Parameter design by a novel method and optimization of the field coil for a 50 kW air-gap HT generator," *IEEE Trans. Appl. Supercond.*, vol. 24, no. 3, Jun. 2014, Art. ID. 5201704.

[135] S. Lee, J. Jeong, K. Jung, J. Hong, and Y. Jo, "Advanced 3-dimensional magnetic field analysis of superconducting machines using analytical method," *IEEE Trans. Appl. Supercond.*, vol. 23, no. 3, Jun. 2013, Art. ID. 4900704.

[136] H. H. Woodson, Z. J. J. Stekly, and E. Halas, "A study of alternators with superconducting field windings: I-Analysis," *IEEE Trans. Power App. Syst.*, vol. PAS-85, no. 3, pp. 264-274, 1966.

[137] A. R. Tariq, C. E. Nino-Baron, and E. G. Strangas, "Iron and magnet losses and torque calculation of interior permanent magnet synchronous machines using magnetic equivalent circuit," *IEEE Trans. Magn.* vol. 46, no. 12, pp. 4073-4080, Dec. 2010.

[138] M. Hsieh, Y. Hsu, "A generalized magnetic circuit modeling approach for design of surface permanent-magnet machines," *IEEE Trans. Ind. Electron.*,vol. 59, no. 2, pp. 779–792, Feb. 2012.

[139] Wen Ouyang, Damir Zarko, and T. A. Lipo, "Permanent magnet machine design practice and optimisation," Presented at 2006 IEEE conference on Industrial Applications, Glasgow, UK, 2006.

[140] A. Zavvos, D. J. Bang, A. Mcdonald, H. Polinder, and M. Mueller, "Structural analysis and optimisation of transverse flux permanent magnet machines for 5 and 10 MW direct drive wind turbines," *Wind Energy*, vol. 15, no. 1, pp. 19-43, Jan. 2012.

[141] A.S. McDonald, "Structural analysis of low speed, high torque electrical generators for direct drive renewable energy converters", Doctor's Dissertation, University of Edinburgh, 2008.

[142] Roark, R.J., and Young, W.C., *Roark's Formulas for Stress and Strain*, McGraw-Hill International Editions, Singapore, 1989, 61h edn.

[143] Juan Pablo Sànchez de Lara García. Wind Turbine Database: Modelling and Analysis with Focus on Upscaling [Master Thesis]. Sweden: Chalmers University of Technology, 2013.

[144] D.C van der Laan, H. J. N. van Eck, B. ten Haken, J. Schwartz, and H. H. J. ten Kate, "Temperature and magnetic dependence of $Bi_2Sr_2Ca_2Cu_3O_x$ tape conductors," *IEEE Trans. Appl. Supercond.*, vol. 11, no. 1, pp. 3345-3348, March 2001.

[145] Y. B. Kim, C. F. Heampstead, and A. R. Strnad, "Critical persistent current in hard superconductors," *Physical Review Letters*, vol. 9, no. 7, pp.306-309, 1962.

[146] F. Grilli, F. Sirois, V. M. R. Zermeño and M. Vojenčiak, "Self-consistent modeling of the I_c of HTS devices: How accurate do models really need to be?," *IEEE Trans. Appl. Supercond*, vol. 24, no. 6, Dec. 2014, Art. ID. 8000508.

[147] Randall. F. Barron. *Cryogenic Heat Transfer*. Ann Arbor, MI, USA: Edwards Brothers, 1999.

[148] L. Quéval, V. M. R. Zermeño, and F. Grilli, "Numerical models for AC loss calculation in large-scale applications of HTS coated conductors," *Supercond. Sci. Technol.*, vol. 29, no. 2, Art. ID. 24007.

[149] V. M. R. Zermeno, A. B. Abrahamsen, N. Mijatovic, B. B. Jensen, and M. P. Sørensen, "Calculation of alternating current losses in stacks and coils made of second generation high temperature superconducting tapes for large scale applications," *J. Appl. Phys.*, vol. 114, no. 17, 2013, Art. ID. 173901.

[150] V. M. R. Zermeno, A. B. Abrahamsen, N. Mijatovic, M. P. Sorensen, B. B. Jensen, and N. F. Pedersen, "Simulation of an HTS Synchronous Superconducting Generator," *Phys. Procedia*, vol. 36, pp. 786-790, Jan. 2012.

[151] H. Polinder, D. Bang, H. Li, and Z. Chen, "Concept report on generator topologies, mechanical & electromagnetic optimisation," Project UpWind, Contract No. 019945(SES6), Dec. 2007. [Online]. Available: http://s3.amazonaws.com/zanran_storage/www.upwind.eu/ContentPages/2473776861.pdf.

[152] K. Zhang et al., "Cold mass support structure design of MRI superconducting magnet developed by IHEP," *IEEE Trans. Appl. Supercond.*, vol. 25, no. 2, Apr. 2015, Art. ID. 4400106.

[153] Cryogenic Technologies Group. Material Properties. [Online]. Available: http://cryogenics.nist.gov/MPropsMAY/materialproperties.htm.

[154] J.R.Bumby, *Superconducting Rotating Electrical Machines.* UK: Oxford University Press, 1984.

[155] Giulio Manfreda, Review of ROXIE's Material Properties Database for Quench Simulation. [Online]. Available: https://espace.cern.ch/roxie/Documentation/Forms /AllItems.aspx

[156] Omid Beik, "An HVDC off-shore wind generation scheme with high voltage hybrid generator," Ph.D. dissertation, McMaster Univ., Hamilton, Ontario, 2016.

[157] Carlo E.D. Riboldi, "Advanced control laws for variable-speed wind turbines and supporting enabling technologies," Ph.D. dissertation, Politecnico di Milano, Milano, 2012.

[158] J. Ekin, *Experimental Techniques for Low-Temperature Measurements: Cryostat Design, Material Properties and Superconductor Critical Current Testing*. Oxford, U.K.: Oxford Univ. Press, 2006.

[159] *Low Level Measurement Handbook*, 7th ed, Keithley. [Online]. Available: http://www.tek.com/sites/tek.com/files/media/document/resources/LowLevelHandbook_7Ed.pdf.

[160] A. R. Kim *et al.*, "Development of critical current measurement system of HTS tape using pulsed current," *IEEE Trans. Appl. Supercond.*, vol. 26, no. 4, Jun. 2016, Art. ID. 9001504.

[161] K. Huang, B. ten Haken, H.H.J. ten Kate, "Critical current of high Tc superconducting Bi2223/Ag tapes," *Physica C: Superconductivity*, vol. 309, no. 3-4, pp. 197-202, December 1998.

[162] M. V. Feigei'man and V. M. Vinokur, "Thermal fluctuations of vortex line, pinning, and creep in high-T, superconductors," *Phys. Rev. B*, vol. 41, no. 13, pp. 8986-8990, 1990.

[163] J. Aponte and H. C. Abache, "Temperature dependence of the critical current in high-Tc superconuctors," *Phys. Rev. B*, vol. 39, no. 4, pp. 2233-2237, 1989.

[164] R. Herzog and R. Wesche and P. Bruzzone, "Field and temperature dependence of critical currents in industrially manufactured high-T c wires and tapes," *J. Phys.: Conf. Ser.*, vol. 97, no. 1, 2008, Art. ID. 012208.

[165] Carmine Senatore, Conductor progress in Eucard2, The 2014 Kyoto Workshop on HTS Magnet Technology for High Energy Physics- The 2nd Woekshop on Accelerator Magnet in HTS (WAMHTS-2). [Online]. Available: https://indico.cern.ch/event/319762/contributions/740335.

[166] E. Pardo and M. Vojenčiak, F. Gömöry and J. Šouc, "Low-magnetic-field dependence and anisotropy of the critical current density in coated conductors," *Supercond. Sci. Technol.*, vol. 24, no. 6, 2011, Art. ID. 065007.

[167] G. Blatter, V. B. Geshkenbein, and A. I. Larkin, "From isotropic to anisotropic superconductors: A scaling approach," *Phys Rev Lett.*, vol. 68, no. 6, pp. 875-878, 1992.

[168] J. H. Durrell, G. Burnell, V. N. Tsaneva, Z. H. Barber, M. G. Blamire, and J. E. Evetts, "Critical currents in vicinal YBa2Cu3O7−δ films," *Phys. Rev. B*, vol. 70, no. 21, 2004, Art. ID. 214508,.

[169] X. Zhang and Z. Zhong and H. S. Ruiz and J. Geng and T. A. Coombs, "General approach for the determination of the magneto-angular dependence of the critical current of YBCO coated conductors," *Supercond. Sci. Technol.*, vol. 30, no. 26, 2016, Art. ID 025010,.

[170] M. Veigel, P. Winzer, J. Richter und M. Doppelbauer, "New FPGA-based and inline-capable measuring method for the identification of magnetic losses in electrical steel," presented at 2015 5th International Electric Drives Production Conference (EDPC), Nuremberg, 2015.

[171] G. Bertotti, "General properties of power losses in soft ferromagnetic materials," *IEEE Trans. Magn.*, vol. 24, no. 1, pp. 621-630, Jan 1988.

[172] A. Krings and J. Soulard, "Overview and comparison of iron loss models for electrical machines," *J. Elect. Eng.*, vol. 10, no. 3, pp. 162-169, Sep. 2010.

[173] F. Fiorillo and A. Novikov, "An improved approach to power losses in magnetic laminations under nonsinusoidal induction waveform," *IEEE Trans. Magn.*, vol. 26, no. 5, pp. 2904-2910, Sep 1990.

[174] D. Kowal, P. Sergeant, L. Dupré and L. Vandenbossche, "Comparison of iron loss models for electrical machines with different frequency domain and time domain methods for excess loss prediction," *IEEE Trans. Magn.*, vol. 51, no. 1, pp. 1-10, Jan. 2015.

[175] V. Zermeño and F. Sirois and M. Takayasu and M. Vojenciak and A. Kario and F. Grilli, "A self-consistent model for estimating the critical current of superconducting devices," *Supercond. Sci. Technol.*, vol. 28,no.8, Jun. 2015, Art. ID. 085004.

[176] D. W. Hazelton, "2G HTS properties beyond critical current," Presented at CHATS-AS Workshop, Cambridge, MA, USA, Oct. 2013.

[177] G. Pellegrino, A. Vagati, P. Guglielmi and B. Boazzo, "Performance comparison between surface-mounted and interior PM motor drives for electric vehicle application," *IEEE Trans. Ind. Electron.*, vol. 59, no. 2, pp. 803-811, Feb. 2012.

[178] Y. Liu, R. Qu, J. Wang, H. Fang, X. Zhang and H. Chen, "Influences of generator parameters on fault current and torque in a large-scale superconducting wind generator," *IEEE Trans. Appl. Supercond.,* vol.25, no. 6, Dec. 2015, Art. ID. 5204309.

[179] Y. Xu, N. Maki and M. Izumi, "Performance comparison of 10-mw wind turbine generators with HTS, copper, and PM excitation," *IEEE Trans. Appl. Supercond.*, vol. 25, no. 6, Dec. 2015, Art. ID. 5204006.

[180] E. Spooner, P. Gordon, J. R. Bumby and C. D. French, "Lightweight ironless-stator PM generators for direct-drive wind turbines," *IEE Proceedings - Electric Power Applications*, vol. 152, no. 1, pp. 17-26, Jan. 2005.

[181] J. Wang, R. Qu, Y. Liu, J. He, Z. Zhe, and H. Fang, "Comparison study of superconducting wind generators with HTS and LTS field windings," *IEEE on Appl. Supercond.*, vol. 25, no. 3, Jun. 2015, Art. ID. 5201806.

[182] X. Song, N. Mijatovic, B. B. Jensen, and J. Holbøll, "Design study of fully superconducting wind turbine generators," *IEEE Trans. Appl. Supercond.*, vol. 25, no. 3, Jun. 2015, Art. ID. 5203605.

[183] Y. Nyanteh, N. Schneider, D. Netter, B. Wei, and P. J. Masson, "Optimization of a 10 MW direct drive HTS generator for minimum levelized cost of energy," *IEEE Trans. Appl. Supercond.,* vol. 25, no. 3, Jun. 2015, Art. ID. 5203504.

[184] Thomas L. Saaty, "Decision making with the analytic hierarchy process," *Int. J. Services Sciences*, vol. 1, no. 1, pp. 83-98, 2008.

[185] H. Polinder and J. G. Slootweg, "Design optimization of a synchronous generatorfor a direct-drive wind turbine," presented at Eur.Wind Energy Conf. Exhibition, Copenhagen, Jul. 2001.

Karlsruher Schriftenreihe zur Supraleitung

Karlsruher Institut für Technologie (KIT) | ISSN 1869-1765

Herausgeber: Prof. Dr.-Ing. M. Noe, Prof. Dr. rer. nat. M. Siegel

Band 001 **Christian Schacherer**
Theoretische und experimentelle Untersuchungen zur Entwicklung supraleitender resistiver Strombegrenzer. 2009
ISBN 978-3-86644-412-6

Band 002 **Alexander Winkler**
Transient behaviour of ITER poloidal field coils. 2011
ISBN 978-3-86644-595-6

Band 003 **André Berger**
Entwicklung supraleitender, strombegrenzender Transformatoren. 2011
ISBN 978-3-86644-637-3

Band 004 **Christoph Kaiser**
High quality Nb/Al-AlOx/Nb Josephson junctions. Technological development and macroscopic quantum experiments. 2011
ISBN 978-3-86644-651-9

Band 005 **Gerd Hammer**
Untersuchung der Eigenschaften von planaren Mikrowellenresonatoren für Kinetic-Inductance Detektoren bei 4,2 K. 2011
ISBN 978-3-86644-715-8

Band 006 **Olaf Mäder**
Simulationen und Experimente zum Stabilitätsverhalten von HTSL-Bandleitern. 2012
ISBN 978-3-86644-868-1

Die Bände sind unter www.ksp.kit.edu als PDF frei verfügbar oder als Druckausgabe bestellbar.

Band 007 **Christian Barth**
High Temperature Superconductor Cable Concepts
for Fusion Magnets. 2013
ISBN 978-3-7315-0065-0

Band 008 **Axel Stockhausen**
Optimization of Hot-Electron Bolometers for THz Radiation. 2013
ISBN 978-3-7315-0066-7

Band 009 **Petra Thoma**
Ultra-fast $YBa_2Cu_3O_{7-x}$ direct detectors for the THz
frequency range. 2013
ISBN 978-3-7315-0070-4

Band 010 **Dagmar Henrich**
Influence of Material and Geometry on the Performance
of Superconducting Nanowire Single-Photon Detectors. 2013
ISBN 978-3-7315-0092-6

Band 011 **Alexander Scheuring**
Ultrabreitbandige Strahlungseinkopplung in THz-Detektoren. 2013
ISBN 978-3-7315-0102-2

Band 012 **Markus Rösch**
Development of lumped element kinetic inductance detectors
for mm-wave astronomy at the IRAM 30 m telescope. 2013
ISBN 978-3-7315-0110-7

Band 013 **Johannes Maximilian Meckbach**
Superconducting Multilayer Technology for Josephson
Devices. 2013
ISBN 978-3-7315-0122-0

Band 014 **Enrico Rizzo**
Simulations for the optimization of High Temperature
Superconductor current leads for nuclear fusion applications. 2014
ISBN 978-3-7315-0132-9

Band 015 **Philipp Krüger**
Optimisation of hysteretic losses in high-temperature superconducting wires. 2014
ISBN 978-3-7315-0185-5

Band 016 **Matthias Hofherr**
Real-time imaging systems for superconducting nanowire single-photon detector arrays. 2014
ISBN 978-3-7315-0229-6

Band 017 **Oliver Näckel**
Development of an Air Coil Superconducting Fault Current Limiter. 2016
ISBN 978-3-7315-0526-6

Band 018 **Christoph M. Bayer**
Characterization of High Temperature Superconductor Cables for Magnet Toroidal Field Coils of the DEMO Fusion Power Plant. 2017
ISBN 978-3-7315-0605-8

Band 019 **Shengnan Zou**
Magnetization of High Temperature Superconducting Trapped-Field Magnets. 2017
ISBN 978-3-7315-0715-4

Band 020 **Ilya Charaev**
Improving the Spectral Bandwidth of Superconducting Nanowire Single-Photon Detectors (SNSPDs). 2018
ISBN 978-3-7315-0745-1

Band 021 **Juliane Raasch**
Electrical-field sensitive $YBa_2Cu_3O_{7-x}$ detectors for real-time monitoring of picosecond THz pulses. 2019
ISBN 978-3-7315-0786-4

Band 022 **Yingzhen Liu**
Design of a superconducting DC wind generator. 2020
ISBN 978-3-7315-0796-3

Die Bände sind unter www.ksp.kit.edu als PDF frei verfügbar oder als Druckausgabe bestellbar.